Duane W. Roller is Professor Emeritus of Classics at The Ohio State University. His previous books include *The Geography of Strabo: An English Translation, with Introduction and Notes* (2014); *Cleopatra: A Biography* (2010); *Through the Pillars of Herakles: Greco-Roman Exploration of the Atlantic* (2005); and *The Building Program of Herod the Great* (1998).

"This book is likely to take its place as the standard introduction for those unfamiliar with the story of ancient geography."
American Journal of Archaeology

"As a general survey and an updated introduction to ancient geography this book is informative and enjoyable reading."
Bryn Mawr Classical Review

"For the first time in several generations, Duane W. Roller offers readers a clear, comprehensive and authoritative survey of ancient geographical thought from its mythic origins in Homer right through to the fall of the Roman Empire. *Ancient Geography* is the distillation of decades of work on the subject by Roller, who is also a distinguished translator of the key books he discusses here. *Ancient Geography* immediately eclipses the introductions to the subject offered by previous scholars and should hold its place as the single key treatment of the topic for generations to come for classicists, geographers and historians alike."
Robert Mayhew, Professor of Historical Geography and Intellectual History, University of Bristol

"In this elegant and readable narrative, Duane W. Roller adroitly recreates the sense of wonder, excitement, and adventure that permeated Greek and Roman geographical initiatives. The result is a vivid tapestry of the many threads of ancient geographical thought that have been untangled from myriad layers of discord, transmission, redaction, and (mis)interpretation in the ancient sources. The book will be warmly and appreciatively welcomed by students of classical history and geography and indeed by anyone with an interest in how antiquity conceived of the world and its features."

Georgia L. Irby, Associate Professor of Classical Studies, The College of William and Mary, Williamsburg

"What Duane W. Roller has achieved in this book is impressive and invaluable. The Greek and Roman grasp of geography, from both spatial and scientific perspectives, developed remarkably over more than half a millennium. So while the approach taken here of explaining this growth chronologically might seem a straightforward task, in fact it is no such thing. Most of the relevant geographical writings and maps are lost. Even some fundamentally important Greek ideas have to be reconstructed from references by later authors who did not always agree with them, let alone perhaps fully understand them. Roller's earlier studies of such giants in this story as Pytheas, Eratosthenes and Strabo make him uniquely qualified to craft an informed, balanced, up-to-date synthesis in defiance of the never-ending obstacles. He writes in a concise, accessible style. Anyone whose imagination is fired by the absorbing puzzle of how the Greeks and Romans envisioned and recorded their surroundings both near and far should read this important book."

Richard J. A. Talbert, William Rand Kenan Jr. Professor of History and Classics, University of North Carolina, editor of *Ancient Perspectives: Maps and Their Place in Mesopotamia, Egypt, Greece, and Rome*

ANCIENT GEOGRAPHY

The Discovery of the World in Classical Greece and Rome

DUANE W. ROLLER

Paperback edition published in 2017 by
I.B.Tauris & Co. Ltd
London • New York
www.ibtauris.com

Hardback edition first published in 2015 by
I.B.Tauris & Co. Ltd

Cover image: The Colossus of Rhodes engraved for the *New Geographical Dictionary*, 1790 (Pictures from History/Bridgeman Images). The Colossus of Rhodes—one of the Seven Wonders of the Ancient World—was a statue of the Greek Titan Helios, erected in the city of Rhodes on the Greek island of Rhodes by Chares of Lindos between 292 and 280 BC. Before its destruction, the Colossus stood over 30 meters (107 feet) high, making it one of the tallest statues of antiquity.

ISBN: 978 1 78453 907 8
eISBN: 978 0 85773 923 0
ePDF: 978 0 85772 566 0

A full CIP record for this book is available from the British Library
A full CIP record is available from the Library of Congress

Library of Congress Catalog Card Number: available

Typeset in Stone Serif by OKS Prepress Services, Chennai, India
Printed and bound by CPI Group (UK) Ltd, Croydon, CR0 4YY

CONTENTS

MAPS

ABBREVIATIONS

AAntHung: Acta antiqua hungarica
AFM: Annali della facoltà di lettere e filosofia, Universitá di Macerata
AJA: American Journal of Archaeology
AncW: Ancient World
ANRW: Aufstieg und Niedergang der Römischen Welt
AntCl: L'antiquité classique
AR: Archaeological Reports
BA: The Barrington Atlas of the Greek and Roman World (ed. Richard J. A. Talbert, Princeton 2000)
BAR-IS: British Archaeological Reports, International Series
BNJ: Brill's New Jacoby
BNP: Brill's New Pauly
BSR: Annual of the British School at Rome
CAH: Cambridge Ancient History
ClMed: Classica et Mediaevalia
CP: Classical Philology
CQ: Classical Quarterly
CR: Classical Review
CRIPEL: Cahiers de recherches de l'Institut de papyrologie et d'égyptologie de Lille
CW: Classical World
EANS: Encyclopedia of the Ancient Natural Scientists (ed. Paul T. Keyser and Georgia L. Irby-Massie, London 2008)
FGrHist: Felix Jacoby, *Die Fragmente der griechischen Historiker* (Leiden 1923–)

FHG: Karl Müller, *Fragmenta historicorum graecorum* (Paris 1841–70)
G&R: Greece and Rome
GRBS: Greek, Roman, and Byzantine Studies
JAOS: Journal of the American Oriental Society
JARCE: *Journal of the American Research Center in Egypt*
JHS: Journal of Hellenic Studies
JRA: Journal of Roman Archaeology
JRS: Journal of Roman Studies
NC: Numismatic Chronicle
OJA: Oxford Journal of Archaeology
OT: Orbis Terrarum
RE: Realencyclopädie der Classichen Altertumswissenschaft (Pauly-Wissowa)
RSA: Rivista storica dell'antichità
TAPA: Transactions of the American Philological Association

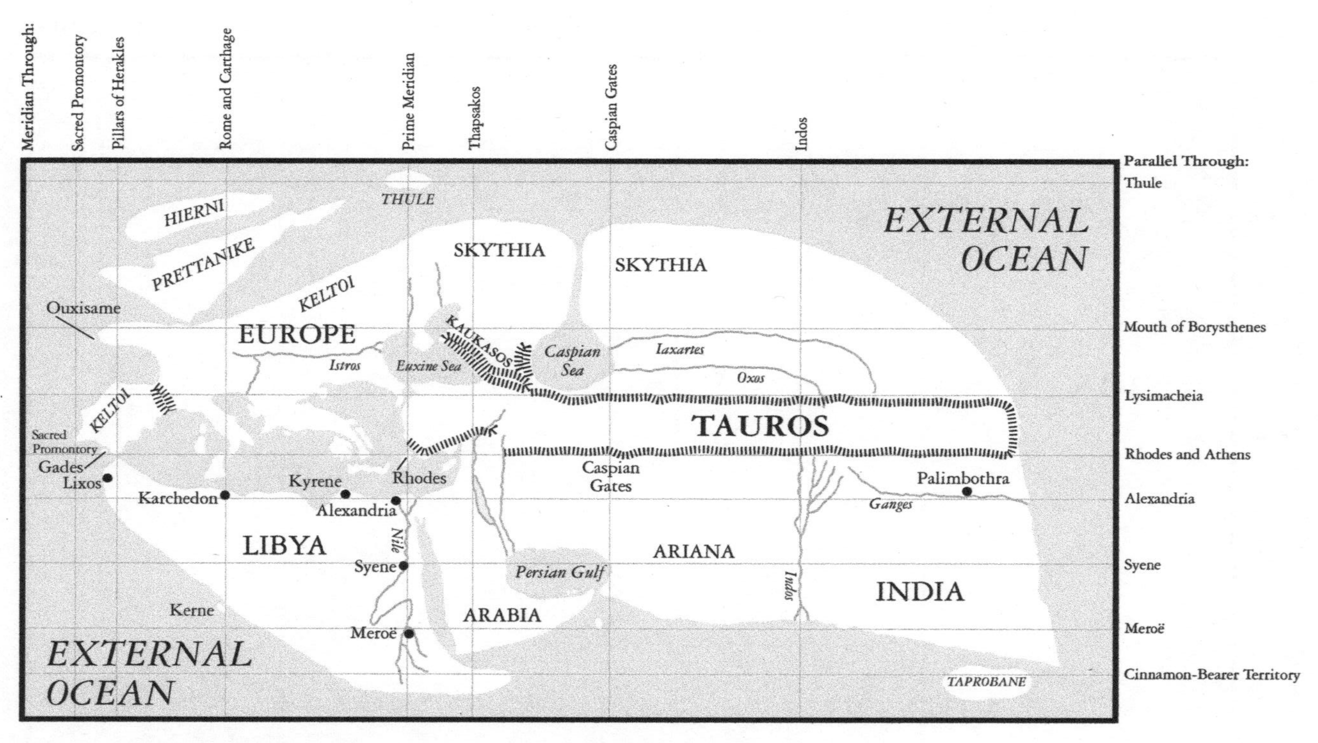

Map 1. The extent of the ancient world in Hellenistic times.

Map 2. The *oikoumene*, as outlined by Eratosthenes.

INTRODUCTION

It is difficult for a person in the twenty-first century, accustomed to maps, aerial photographs, and instant access to views of any place in the world, to comprehend the astonishing feats of ancient travelers and geographers. Greek scholars created a theoretical construct of the earth as a whole, and sailors, merchants, traders, and military commanders wandered far from home and brought back the topographical and ethnographic data that became the basis of geographical thought. Until the advent of the railroad in the mid-nineteenth century, the only means of transportation was by foot, beast of burden, or ship, limited to a few miles a day. One's horizon was literally that: the short distance one could see. The major tool for determining location was the eyes, and going several days from a known place could put the traveler in unfamiliar territory, which could only be related to the starting point by backtracking, in so far as that was possible. Northerly and southerly movement could be determined by changes in the height of celestial bodies, but there was no way to calculate easterly or westerly progress beyond the dead reckoning of distance traveled.

With such limitations, it is phenomenal that so much was actually learned about geography in Greco-Roman antiquity. Perhaps assisted by the enclosed nature of the Mediterranean and Black Sea system, which provided a perimeter to explore, by the sixth century BC Greeks had covered essentially all of its extent, and had begun to realize that the earth was a very large sphere (its curved surface was apparent to any sailor). In the following centuries, there was penetration as far as the arctic North Atlantic, south along the African coast (and perhaps all the way around the continent), and east to India, with some knowledge of the Chinese world beyond. By the Roman period all the eastern hemisphere was known except for Siberia and interior southern Africa, although the level of comprehension varied greatly in different areas. A parallel theoretical evolution had determined the size of the earth and the placement of its known portion on its sphere, with the understanding this was a very small part of its entirety. Moreover, there was discussion of the changes in climate and the length of day as one went north or south. It was even suggested, since the inhabited world was so small, that there might be other continents, a New World beyond the familiar one. Geography as a scholarly discipline developed in the latter third century BC and brought the empirical and theoretical data together. By the second century AD all the then known world had been mapped.

The study of ancient geography relies on three components, often separate but which must be considered collectively. First is the practical information: the topographical data gathered by the people who actually traveled. This phase of geography began with the first person who floated on a log to a neighboring island, or who went over the mountains to a nearby village, thereby expanding the geographical horizon. From late Neolithic times the most efficient way to travel was by ship, so sailors were the first long-distance voyagers, and they recorded in their minds sailing directions, winds, harbors, and the peoples that they encountered. These vagaries of seamanship were well known to Odysseus—the first sailor in the Greek world to emerge as an individual personality—and he was more concerned about the people whom he came upon (who were usually dangerous) than the places. Eventually a repertory of sailing itineraries and distances evolved, recorded in days of travel, information that was exchanged and shared in seaport cities. Land travel was slower to develop because of the maritime nature and rugged topography of much of the Greek world, but there were places that had no accessible coasts, such as interior Anatolia, and here primitive paths and roads were the first routes of travel.

The second element of ancient geography is the creation of a theoretical structure about the nature of the world. This was originally the province of the natural philosophers, as empirical evidence was lacking beyond one's immediate surroundings and the visibly curved surface of the earth. Moreover, early sailors and other travelers had no interest in theory. It was probably the Pythagoreans, perhaps as early as the sixth century BC, reinforced by Plato and others, who suggested that the earth was a sphere, and by the third century BC it was possible to calculate its size—using astronomical data—and to locate known places on its vast surface. There was also speculation about the history of the earth and the changes that it had undergone, as well as its climate. These various disciplines—natural philosophy, astronomy, and geology, as well as empirical data—came together in the latter third century BC when Eratosthenes of Kyrene published his *Geography*, the first work devoted to the topic and, incidentally, the first use of the term. Eratosthenes determined the size of the earth and then located known places on it, largely through relative positioning based on data from travelers' reports. He realized the inherent flaw in this technique, and, by the following century, trigonometric methods were used to position places astronomically, and therefore accurately, but this was never fully utilized in ancient times due to a lack of qualified observers. To the end of antiquity, most places were still located using on-the-ground reports from travelers.

The Romans continued exploration—but generally for military purposes—and learned about parts of the inhabited world that had previously been unknown, such as northern Europe. On the other hand, there were few further developments in geographical theory, other than a continued interest in issues regarding mapping of the earth, especially the problem of projecting a curved surface onto a flat map. But the geographical creativity of antiquity effectively ended in the middle of the second century AD with the *Geographical Guide* of Ptolemy of Alexandria. After this time there was continued refinement of detail, both theoretical and topographical, yet with the advent of Christianity, geography turned toward being driven almost solely by biblical exegesis.

The third consideration for the modern student of ancient geography is the sources. The land is still there, relatively unchanged, but the loss of ancient literature has fallen especially hard on geographical texts. Nevertheless literature remains the primary source for understanding ancient geography, with archaeology playing a secondary role, though especially valuable for physical evidence of the spread of Greek and Roman culture into remote areas. Only four geographical handbooks are extant from antiquity, a poor showing of the nearly 250 known Greek and Roman geographers,[1] most of whom are known solely through quotations from later authors, or by name alone. Of the four that are extant, all are from the Roman period. The earliest, and most important, is the *Geography* of Strabo of Amaseia, written in Greek and 17 books in length, completed in the early first century AD. Without this extensive and complex work almost nothing would have survived about the history of geography before the Roman period: major geographical authors whose texts are lost, including Eratosthenes, the inventor of the discipline, are known almost entirely through Strabo's summaries of their work. This of course creates a problem, since Strabo had his own agenda and was writing about geographical theory through the perspective of an educated Greek functioning in the Roman empire. Nevertheless, as can be seen in the pages that follow, Strabo is the primary source for the topic of ancient geography.

The handbook written by Pomponius Mela in the 40s AD, titled *Chorographia*, has the unique status of being the only extant free-standing geographical treatise from antiquity written in Latin, and is important for that reason, but its brevity—only three books long—means that it is less significant than the other surviving writings. More important are the geographical books (2–6) of the *Natural History* of Pliny the Elder, written a generation after Mela. Though part of a much longer work, Pliny's five geographical books could stand alone as an independent treatise, the fullest surviving geographical account in Latin. He made exhaustive use of material obtained in the early years of the Roman empire, including a dazzling array of toponyms.

In the mid-second century AD, Ptolemy of Alexandria completed his *Geographical Guide*, a work of great importance in medieval and Renaissance times. It is effectively the terminal point of the present study. Ptolemy was a mathematician and astronomer, and his primary concerns were the role of the latter in geography and how to make a map of the known world. In the process he located (with varying degrees of accuracy) over 8,000 places, from the Baltic south to the legendary Mountains of the Moon in central Africa, and east to the Malay Peninsula. This is the most complete topographical list from antiquity, and represents the culmination of ancient geographical knowledge.

Needless to say, these four works are not the totality of extant ancient literature on the topic of geography. Many—if not most—surviving ancient authors from Homer on have geographical material, including numerous writers who are not usually thought to be geographical in their orientation, such as Aeschylus and Vergil. Polybios and Julius Caesar, hardly remembered as geographers, were skilled practitioners of the discipline. There is also a unique geographical literary genre, the *periplous* or the coastal sailing manual. Some of these actually survive, and information from others lies buried in extant geographical works. Geography also has its literary aspects, and its data can be used metaphorically and allegorically, or even to create a fantasy world.[2] But one returns to Strabo again and again to learn about the practical aspects of ancient geography.

The modern study of ancient geography began in the Renaissance, when the major texts became available and were of great use to the explorers of the era: Columbus was totally familiar with the extant writings on the topic. Yet like so much of classical studies not considered to be in the literary mainstream, examination of the topic has languished in modern times until recently. In English, the seminal work remains E. H. Bunbury's *A History of Ancient Geography* (London 1883). There are also J. Oliver Thomson's *History of Ancient Geography* (Cambridge 1948), and *The Ancient Explorers*, by M. Cary and E. H. Warmington (revised edition, Baltimore 1963). Despite the age of these works, all are still important in differing ways.

It is only in the last two decades that English editions of the four major geographical handbooks have begun to appear, beginning with Pomponius Mela (F. E. Romer, *Pomponius Mela's Description of the World*, Ann Arbor 1998), and followed by Ptolemy (J. Lennart Berggren and Alexander Jones, *Ptolemy's Geography: An Annotated Translation of the Theoretical Chapters*, Princeton 2000), although this edition does not include the topographical material (at present only available in German: Ptolemaios, *Handbuch der Geographie*, ed. Alfred Stückelberger and Gerd Grasshoff, Basel 2006). A new translation of Strabo's *Geography* has recently appeared (Duane W. Roller, *The Geography of Strabo*, Cambridge 2014); Pliny's geographical chapters await their English editor. Needless to say, there are numerous other recent articles and books about various aspects of ancient geography, many of which appear in the bibliography.

A Note on Ancient Measurements

At first, distances were recorded in sailing days and camel (or other beast of burden) days. Eventually various populations in the ancient world developed their own measurements: the Greeks the stadion, the Persians the parasang, the Egyptians the schoinos, and the Romans the mile. All of these were used in Greek and Roman geographical documents. Except for the Roman mile (about 4875 feet), these measurements were highly variable and uncertain, and often based on intuitive or traditional data rather than accurate measurement. It was only in the Roman period that standard distances were established, with the invention of the odometer, perhaps by Roman military engineers.[3] Yet calculations still remained rough, and thus modern equivalents are avoided. Very approximately, there are about 30 stadia to the parasang and 8–10 stadia to the mile, but any attempt at accurate translation into modern lengths is impossible as well as misleading.

Acknowledgements

Among the many who assisted in the preparation of this work are Georgia L. Irby, Molly Ayn Jones-Lewis, Paul T. Keyser, Letitia K. Roller, Richard Stoneman, Richard Talbert, Lisbet Thoresen, Alex Wright, Sara Magness, and many others at I.B.Tauris, the Harvard College Library and the libraries of the Ohio State University (especially its interlibrary loan services) and Stanford University.

CHAPTER 1

THE BEGINNINGS

The Mediterranean—which surrounds the Greek heartland—is well suited for seafaring and exploration. Rarely is one out of sight of land. Its largest expanse is southeast of Italy and Sicily, where a sail from Syracuse to Kyrene might briefly be as much as 200 miles from shore, but a deviation to the east would keep land in sight.[1] Rugged coastlines can be seen far away: Mt Etna on Sicily is visible 100 miles out to sea. All the islands in the Mediterranean can be seen from some other point of land, so every locality could be explored without being in the open sea, and in many places one could cross the Mediterranean and always be in view of land.

Seamanship in the Greek world began in the Middle Neolithic Period. At Sesklo in northeastern Thessaly (not far from where Jason and the Argonauts were to set sail), obsidian from the island of Melos has been found in contexts from as early as the fifth millenium BC,[2] and it is hard to imagine that this obsidian made its journey by any means other than ship, although contemporary ships may have been little more than primitive rafts, much like the one Odysseus used to escape the affections of Kalypso, and whose construction Homer described in detail.[3] The journey from Melos to Sesklo need not have been entirely by sea: one could go from one western Cycladic island to another and reach the coast of Attika, and then continue overland, but some degree of sea passage was required. The adventurous person who first brought Melian obsidian to the mainland unwittingly inaugurated the era of Mediterranean exploration and thus began the path to the discipline of geography. By the latter third millenium BC ships were represented in art: on a terracotta "frying pan" of *c.*2500–2100 BC, probably from the island of Syros (not far from Melos), is an incised drawing of a low, long vessel with a high stern and 16 oars on each side, the earliest view of a ship from the Mediterranean world.[4] Somewhat later, at the opening of the Middle Bronze Age, seafaring in the Aegean was well advanced.

To be sure, the Aegean peoples came late to the exploration of the Mediterranean. Egyptians had long sailed from the Nile Delta to the Levant, and knew about Crete and southwestern Anatolia. Assyrians may have ventured west into the Mediterranean.[5] The Greeks themselves knew some details about the Minoans on Crete, who covered much of the eastern Mediterranean and perhaps went as far as Italy, as several toponyms such as "Minoa" suggest, although it is also possible that these names may be from a later period and a supposed history of a Cretan presence.[6] The Minoan thalassocracy, so much a part of the lore of the early Aegean, and the tale of the death of King Minos in Sicily, are not documented before the fifth century BC, but this may be due to the nature of the surviving evidence, and there was an ancient "Tomb of Minos," an impressive multi-story structure, in southern Sicily many generations previous to this time.[7] These voyages—Egyptian, Cretan, and Assyrian—mean that before the Trojan War the sailing routes across the eastern Mediterranean had been well established. Travels for commercial, political, and military reasons created a database of topographical knowledge that could be used for geographical purposes.

The Argonauts

The first great voyage of the Greek world was that of Jason and the Argonauts, who sailed in the *Argo* to far-off Colchis, in search of the Golden Fleece. The myth is placed in the generation before the Trojan War (thus somewhat after 1300

BC), since the brothers Peleus and Telamon, the fathers of Achilles and Aias, participated, although the list of Argonauts is extremely long and such genealogical associations are always dubious. Understanding the story of the Argonauts is difficult because of the minimal early source material: the most familiar version, the *Argonautika* of Apollonios of Rhodes, is from the third century BC and reflects contemporary geographical knowledge unlikely to have been in the original account.[8] Yet the tale is early and may contain elements from the very beginnings of seamanship. The unusual properties of the *Argo*—it could speak because it had wood from the sacred oak at the oracle of Dodona built into it—go back to an early era when seamanship and navigation still seemed to be a matter of magic rather than skill.[9]

Deconstructing the Argonaut story is problematic. One would think that it would have been a favorite topic of discussion at Troy, given that sons of the adventurers were present and prominent, but the tale is not mentioned in the *Iliad*, and Jason is named merely because his son Euneos, a minor character in the epic, brought wine from Lemnos for the Achaians.[10] It is only in the *Odyssey* that there is any hint of the expedition, in the sailing instructions given to Odysseus by Kirke. She tells him that only one ship has passed through the Clashing Rocks (Planktai): the *Argo*, with Jason aboard.[11] Kirke was the sister of Aietes, the father of Medea, and although she never speaks of the relationship between her niece and Jason, the story has elements of a family tale. The Planktai, also called the Wandering Rocks, were normally located at the entrance to the Black Sea,[12] yet it is by no means certain that Homer had placed them there. Thus Homer's knowledge of Jason and the Argonauts was of a great expedition that overcame at least one major peril and which may have entered the Black Sea and perhaps was associated with Aietes. Homer did not specify where Aietes lived, or reveal any obvious knowledge of the Black Sea. Hesiod, somewhat later, had more detail, including the first citation of the Phasis River (modern Rioni, at the southeast corner of the Black Sea), the river that flows through Colchis. He was also the first to give the name of Aietes' daughter, Medea, whom Jason took back home to Iolkos.[13] The connection between the Argonauts and Colchis is also documented by the Corinthian poet Eumelos, a rough contemporary of Hesiod.[14] In his version, Aietes, a local ruler in the Corinthia, went to Colchis and established himself there; for this reason Jason eventually ended up in Corinth with Medea, events that led to the dismal conclusion recounted in Euripides' *Medea*. Thus the entire Argonaut tale and its wide-ranging settings were in place by the eighth century BC—except perhaps the actual reason for the expedition—although details continued to be refined into Hellenistic times.

The story follows a series of familiar mythological formulae: the deposed heir to the throne who inconviently returns to claim his inheritance, the sending of him on a expedition which he is not expected to survive, the liaison with a foreign princess, and the unexpected return home. In mythology, it is yet another account of the dysfunctional aristocratic family of the Greek Bronze Age. But one of its most familiar elements is not mentioned in these early versions, and may come from a totally different source. The Golden Fleece was known to Hesiod only in the context of another early tale of exploration: that of Phrixos and Helle.[15] When the two stories were joined is not certain, but eventually the goal of the Argonauts was the fleece itself, one of the more interesting items of Greek mythology. Despite the magical origins, it represents two essential elements of wealth in early Greek society: sheep and gold. It remained a powerful token: as late as the first century AD the grove in Colchis where it hung was still visible.[16] There were many ways to rationalize the story—perhaps the most interesting is that fleeces were used to wash gold in the Caucasus, a practice still followed in the first century BC—but it is perhaps nothing more than a striking metaphor for Greek attempts to capitalize on the wealth at the far end of the Black Sea.[17] The Greeks were beginning to explore that sea at the time of Hesiod and Eumelos, their first attempt to move beyond the Mediterranean proper into other waters. It was a dangerous area that was called Somber or Black, guarded by the infamous Wandering or Clashing Rocks. The sea was not named in extant literature until the fifth century BC,[18] but Hesiod nonetheless knew two of the rivers that emptied into it, the Istros (modern Danube) and the Phasis (modern Rioni), the latter at the farthest point of the sea in the region called Colchis. To go from the Aegean to the mouth of the Phasis would exactly replicate the voyage of the *Argo*, and it seems probable that by the time of Eumelos and Hesiod this was done on a regular basis—a journey of more than 1,000 miles from Iolkos, where the *Argo* started. Eventually there was a homonymous Milesian settlement at the outlet of the Phasis, probably established in the fifth century BC.[19] But Hesiod knew of the river long before Greeks settled at its mouth, and the Argonautic expedition came to be explained as a search for precious metals[20]—simplistic but perhaps the original reason for curiosity about the farther areas of the Black Sea, an interest perhaps going back to the Bronze Age. Gold mining had long existed in this region, not so much around the Black Sea coast but in areas of the Caucasus reached from the upper Phasis, especially the district known in antiquity as Iberia. Moreover the river was the western end of a trade route that, while difficult, extended far into the interior.[21]

The voyage of the *Argo* is a mythological remnant of the first long trading voyages made by Greeks. The tale that is familiar today evolved over many centuries and from several different sources, intermingling Thessalian, Corinthian, and Italian elements. Some of it is exceedingly old, perhaps pre-Greek, and other parts are as recent as the emergent Greek knowledge of the Black Sea in the Archaic period. Perhaps the data provided by the early expedition of the Argonauts stimulated Greek exploration of the southeastern Black Sea in later times, and by the eighth century BC some adventurous Greeks, knowing about the Argonauts, decided to see if the wealth of Colchis really existed. Greeks were at the mouth of the Phasis before 700 BC, although no permanent settlement was established at that time.[22] These traders and merchants were not geographers, but they brought back toponyms and sailing directions, and a sense of a world beyond the Mediterranean.

The Catalogue of Ships in the *Iliad*

In the second book of the *Iliad* is a well-known list of the Achaian forces who marshalled at Troy, followed by a shorter one of the Trojans.[23] It has long been recognized that these catalogues are, in many ways, independent of the narrative of the *Iliad*—there is remarkably little correspondence, especially in the Trojan catalogue, between those listed and the action of the poem—and they probably represent a different, later, and more geographical tradition than the bulk of the *Iliad*. Some of the places mentioned do not seem to have been occupied until long after the Trojan War era.[24] Regardless of its origin, the Achaian catalogue is the earliest geographical document in Greek literature, with 175 toponyms cited.[25] It was probably Central Greek in origin, since the Boiotians are presented first and located in 30 places, but have little role in the action of the epic: in fact, more than a quarter of the names are from Boiotia and adjoining areas. The catalogue is like a map, and is selective in ways that will never be understood. The presentation of topographical and ethnographic details—often as little as a single word or less than a line of poetry—established the pattern of topographical description in Greek geographical writers, who regularly deconstructed the catalogue in detail. In the Hellenistic period, entire treatises were devoted to this, most notably that of Apollodoros of Athens, who in the second century BC wrote 12 books on the catalogue, and his contemporary Demetrios of Skepsis, who produced 30 books on the Trojan one. Neither of these works is extant, but Strabo made extensive use of both and added many comments of his own.

The Return of the Heroes From Troy

Somewhat easier to interpret in the context of developing geographical knowledge is the cultural event known to the Greeks as the *Nostoi* or the Returns. This refers to the homecoming of the Achaians who had survived the Trojan War, representing a transition to the post-war era in the early twelfth century BC. Those who had no difficulty and ended their lives back home, such as Nestor and even Agamemnon, are of no interest geographically, but many others had wide-ranging wanderings into regions previously unknown. Like much of the material associated with the war itself, details evolved and were enhanced over time, especially as geographical knowledge expanded, with the story of Aineias, or Aeneas, perhaps the most obvious example. Moreover, there is always the impossible task of accurately separating the various layers of material: the event itself, which may have occurred in the immediate post-war era, the earliest written records of the event, which would be no earlier than the time of Homer and Hesiod, hundreds of years after the war, and the later evolution of the tale all the way into the Hellenistic and Roman periods.[26]

There are many stories of wanderings but little physical evidence, and the stories seem to have become richer as time passed. Often there is no evidence before the Hellenistic period, which is not cause for outright rejection of an account but a reason to be cautious. A case in point is the journey of Philoktetes, the Thessalian hero who was left behind on Lemnos before the war but without whom Troy could not be taken, and who eventually arrived late in the conflict. According to Homer he returned home without difficulty[27] but later it was believed that he was eventually forced to leave Thessaly, moving to southern Italy and founding Petelia (at modern Strongoli in Calabria). The earliest extant account of this tale is not before the second century BC,[28] at a time when Greek towns outside the heartland were particularly interested in their heritage and any possible connection to the world of Troy. This is not enough to dismiss the story but it creates a difficulty in relating Hellenistic tales of city foundation to actual events of the end of the Bronze Age. Nevertheless, there may have been faint memories of early history that were still available in much later times.

Many post-war refugees moved within the Eastern Mediterranean, such as to the Levant and Egypt, but these were regions already known to the Greeks and did not represent the discovery of new places. Italy, however, was the land of opportunity in this era. The first permanent Greek settlement in Italy, on the island of Pithekoussai (modern Ischia) was not founded until the early eighth century BC.[29] Yet much earlier there had been voyages to Italy by the Mycenaean Greeks who were living on the western coast of the Greek peninsula and the adjacent islands. These would not have been difficult journeys, as Italy is visible from the northwestern Greek coast. Many cities in southern Italy, like Petelia, claimed to have been founded by those who fought at Troy or their immediate descendants. Metapontion was said to have been settled by associates of Nestor from Pylos.[30] Diomedes of Argos was believed to have been one of the widest wanderers. He returned home without difficulty,[31] but found that his wife, Aigialeia, had made other plans in his absence, and quickly moved on, avoiding the fate of his neighbor Agamemnon. He ended up on the eastern coast of Italy and was credited with founding a number of cities along the entire length of the Adriatic. Eventually he was worshipped as a god[32] and, by some accounts, he and his companions were turned into the birds that were still visible in Roman times on the Adriatic islands named after him: the modern Isole Tremiti off the Apulian coast, where his tomb was located.[33] The story of Diomedes is a mixture of post-war wandering, city foundation, and local hero and divinity cult. As usual, much of the detail is not extant before the Hellenistic period, but the tale preserves a memory of the earliest Greek contact with the Adriatic, whose long narrow shape encouraged exploration as far as its northern end in the land of the Enetians, or Venetians, where there was a sanctuary to Diomedes as late as Roman times.[34]

Archaeological evidence to support an early date for these settlements is scant but nonetheless present. The major Mycenaean site in southern Italy is Scoglio di Tonno, at the northwestern edge of modern Taranto, where material of the Late Helladic IIIc period (the twelfth century BC or the immediate post-war era) has been found.[35] In fact, Mycenaean pottery has been discovered in southern Italy and Sicily as early as the Late Helladic II period (roughly the fifteenth century BC), indicating trade contacts with the Aegean world from an early date.[36] Thus even before the Trojan War, southern Italy was vaguely part of the Greek horizon, and the migrations and disruptions of the post-war era consolidated this contact.

The wanderings of Odysseus deserve special comment, due to their extent and the early source material preserved in the *Odyssey*. This is not the place to discuss the complex and undetermined relationship between the character Odysseus, who fought at Troy and had problems returning home, and the material preserved in the *Odyssey*, composed in its present form hundreds of years later, much of it no doubt gathered as its author, Homer, sat in a waterfront taverna and heard stories from sailors who had returned from remote places with strange tales, as happens even today in any port city. In addition, there are other layers of the data, including details about Odysseus not recorded by Homer, and the inevitable evolution of the myth from Homeric into Hellenistic and Roman times.[37] Nevertheless, the *Odyssey* is examined here as the story was presented by Homer, with a view toward gleaning whatever information it may reveal regarding early travels in the post-war era.

Unlike the later accounts of sailing or actual journeys between specific points, the *Odyssey* is a series of disconnected episodes, the type of encounters that have been typical of sailors' tales from ancient to modern times: hostile coastal peoples, navigational hazards, and dangerously seductive women, all refined and exaggerated through retelling. As a work of literature it begins a tradition that remains vigorous today in modern science fiction, based on the idea that once one leaves the known world and enters the unknown, the rules of society and the laws of physics change drastically.

Since Homer's day there have been unending attempts to locate the places that Odysseus visited, to some extent a futile exercise because the *Odyssey* is a work of literature not the log of a journey. Nevertheless the account reflects some knowledge of remote regions and peoples—from perhaps as early as the thirteenth century BC, and certainly from Homer's day of several hundred years later—and as such is an important document in the development of Greek geographical knowledge.

After the fall of Troy, Odysseus and his surviving companions headed home. They were not in so much of a hurry that they were unable to indulge in some pillaging on the way, sacking Ismaros in Thrace, which was the city of the Kikonians, who had been Trojan allies.[38] Ismaros is an actual place: although it has not been specifically located, it lies within a limited area on the north Aegean coast near modern Komotini, where in later times there were still tokens of Odysseus' passage, such as an Odysseia River and a shrine to Maron, a local resident who gave him the wine that he in turn was to give Polyphemos. The Kikonians survived in this region into at least the fifth century BC.[39] When Odysseus left them, a great storm blew him and his men for nine days, and they were unable to round Kythera at the southern end of the Peloponnesos—the last-known toponym mentioned—arriving on the tenth among the Lotus Eaters.

The land of the Lotus Eaters implies the languid north African coast, a reasonable place to end up if one could not make it around the Peloponnesos, and it has been located there since at least the fifth century BC. In Hellenistic times opinion centered on the island of Meninx (today the Tunisian resort of Jerba), a suggestion originating with Eratosthenes.[40] The lotus in question is probably a variety of water lily.[41]

After the Lotus Eaters, the encounters become more fantastic and less real. The Kyklopes are described in detail, with topographical and cultural elements that imply an actual place, although within the formula of the folk tale of a paradise that turns into horror, the ideal society with irredeemable flaws.[42] There are many groups of Kyklopes known in Greek literature, who do not seem particularly to relate to one another: later accounts connected those that Odysseus saw with Hephaistos and placed them in the Lipari Islands north of Sicily, which, given the next encounter, has some plausibility.[43]

Aiolia also suggests the Lipari Islands, which have been called the Aiolian Islands since before the fifth century BC: it was believed that the winds implicit in the Homeric tale reflected the effects of the volcanic phenomena that are still active in the islands.[44] The Laistrygonians have been located around Mt Etna since the time of Hesiod.[45] Next was the home of the sorceress Kirke, Medea's aunt,[46] generally placed on the Italian coast where it makes its sharp turn to the north between Naples and Rome, a locality still called Monte Circeo. Of particular interest is that this was a region of unusual herbs and roots, something that excited the curiosity of Theophrastos.[47] A temple to Kirke was there in Roman times, where tourists were shown a golden cup which may have been the one that she used to serve her potion to Odysseus.[48]

After a diversion to the underworld, Odysseus and his companions continued to other less mythical places. The Sirens, who lived on an island, are represented by a number of toponyms around the Bay of Naples, most notably Surrentum (modern Sorrento), and were still honored in the Roman period.[49] Skylla and Charybdis are almost without exception placed in the Straits of Messina.[50] Skylla, the monster, was said to be on the mainland, and Charybdis, the whirlpool, survives (much reduced due to tectonic activity) just off modern Messina. Thrinakia, where the cattle of Helios were encountered, was the ancient name (as Trinakria) for Sicily.[51] Eventually Odysseus landed alone on the island of Kalypso, Ogygia, where he spent many years. It has been located on several islands in the Mediterranean but, for reasons outlined below, it must be somewhere around Sicily or Italy, and Kallimachos' suggestion of the Maltese island of Gozo is perhaps the most plausible and is also believed by the locals.[52]

From Kalypso's embraces Odysseus returned home, after visiting Scheria. Kalypso gave him sailing directions from her home to Ithaka:

> He looked upon the Pleiades, and late-setting Boötes, and the Bear, which is also called the Wagon, circling in a fixed place, watching Orion, and which alone has no share in the baths of Ocean. Kalypso, the heavenly goddess, had told him to have it on his left hand as he sailed on the sea.[53]

In other words, he was to sail east to reach Ithaka. This is perhaps one of the most significant geographical passages in the *Odyssey*, demonstrating without question that some of the poem is grounded in navigational details available to Homer. In addition to the Bear, the three other constellations—the Pleiades, Boötes, and Orion—were also of special interest to sailors, but not of particular use to Odysseus at this point. Yet this is the only time that constellations are mentioned anywhere in the *Odyssey*, and the passage demonstrates that Homer had access to details of the sailing route between western Greece and Italy or Sicily. Thus Ogygia must be placed somewhere around Italy or Sicily, and the accumulation in approximately the same region of so many of the other points visited by Odysseus becomes highly plausible.

To be sure, arguments about the location of the wanderings become dangerously circular, but the explicit sailing directions from Ogygia to Ithaka provide the key, and at least show that some of the places Odysseus was believed to have visited were in that region of opportunity in both his and Homer's day: southern Italy and Sicily. The profusion of names in this area—Kalypso's island, Thrinakia, Aiolia, Kirke's home, Skylla and Charybdis, the Sirens, and the Laistrygonians—may be due to later elaboration, but only in part. Homer knew about this area, however vaguely, and wove it into the fabric of the *Odyssey*. Yet how this information might have been divided between the knowledge of Homer's own era and information going back to the world of Odysseus cannot be determined.

The World of Homer (Map 3)

It seems that in Homer's time Greeks were aware of the Adriatic coast of Italy—perhaps its entire length—and the west coast almost as far north as the mouth of the Tiber, and also Sicily and part of the north African coast west of Egypt. It remains to determine whether Homer knew about a wider world. Two sparse comments in the *Odyssey* have been interpreted to mean that he had some sense of the astronomical phenomena of high latitudes. The Laistrygonians lived where day and

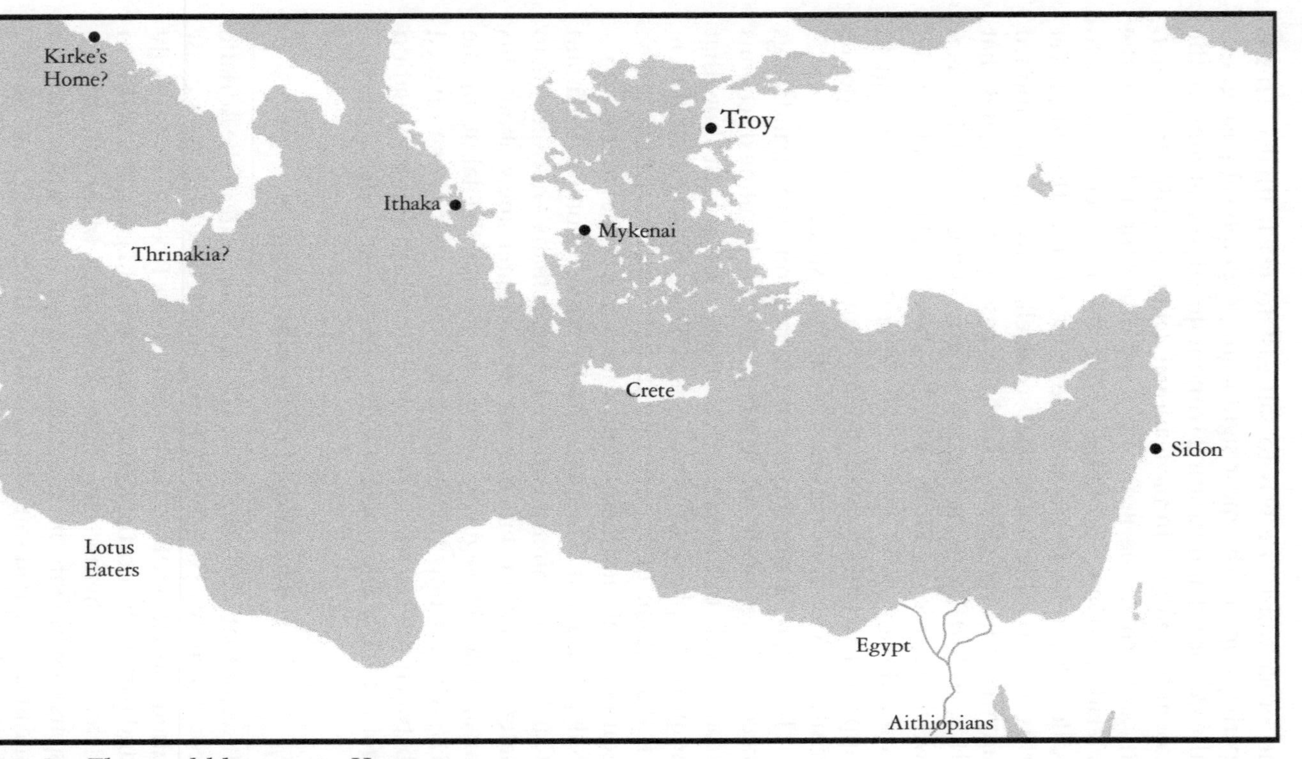

Map 3. The world known to Homer.

night were close together, an unusual statement that the Homeric critic and geographer of the second century BC, Krates of Mallos, took to be a reference to the midnight sun of the Arctic.[54] This seems improbable, as the Homeric line is in the midst of comments on the pasturing habits of the Laistrygonians, and probably refers to the relationship of the work of their shepherds to the day and night.[55] On the other hand, the Kimmerians lived in a region of eternal night,[56] perhaps a more plausible suggestion of the northern winter. It is possible that some of Homer's informants in the hypothetical waterfront taverna had traveled to the north, or knew about the trade routes to the Baltic that had been in place since the Bronze Age. Yet the reference may merely be more a meteorological statement than an astronomical one, since the Kimmerians lived adjoining the Ocean and the "night" is said to be "mist and cloud."

The Ocean itself was mentioned more than 30 times by Homer. It encircles the inhabited earth, as depicted on the Shield of Achilles, and the Aithiopians live in its vicinity.[57] The Ocean has a consistent sense of remoteness, something beyond the normal reaches of human contact. Nearly half the citations describe it as a river or stream, and it has a mouth.[58] But it is an unusual river, for it is "deep eddying," "deep flowing," and, most interestingly, "backward flowing."[59] It seems that Homer integrated two views of the Ocean, presenting it either as a conventional river with a mouth at the remote limits of the world, or a unique body of water that encircles the earth.[60] Its backward-flowing quality may be a faint suggestion of the tides. The word "Ocean" is not Greek—Minoan has been suggested—and its discovery was attributed in antiquity to the Phoenicians.[61] To early Greeks, there were two places that the Ocean could be accessed: the Red Sea and the Atlantic beyond the Pillars of Herakles. Egyptians had known about the Red Sea and the adjacent Indian Ocean from earliest times,[62] and Homer was knowledgeable about Egypt as far up the Nile as Thebes, as well as the Aithiopians beyond.[63] He may have learned about the Ocean from the same source that he heard about the Aithiopians, but it is more likely that his information came from Levantine informants. The Phoenicians had been through the Pillars of Herakles and into the Ocean before Homer's time,[64] and he knew about their town of Sidon and its famous artisans. A Sidonian silver krater was one of the prizes in the games for Patroklos, and Hekabe had fine clothing from Sidon, a city that Alexandros/Paris had visited on the journey that led to Helen.[65] It seems that by Homer's day, if not well before, there was regular contact between the Aegean and the Phoenician heartland, and this was where Greeks first learned about the External Ocean.

Although Homer's knowledge of the Ocean was vague and contradictory, he did work it into his mythological world, most notably as the location of the Elysian Plain. This was on the Ocean and had an ideal climate, an earthly paradise where Menelaos and Helen would end up. The concept of the Elysian Plain is probably pre-Greek: a mythical place that could not be precisely located.[66] Nevertheless, Homer's seeming knowledge of sites in or on the Ocean eventually gave rise to the idea that his world was far wider than it was, and, as understanding of the western Mediterranean expanded, Homer's world moved with it, eventually beyond and into the Ocean. In time, Kalypso's island of Ogygia was said to be far out in the Atlantic,[67] and inventive use of toponyms created an entire Iberian episode to Odysseus' wanderings.[68] Yet it is highly improbable that Homer knew about anything west of Sicily and southern Italy. To him the Ocean was merely a great and unusual river flowing around the extremities of the earth, something that he learned from the Phoenicians and which would become a pervasive geographical theory.

Nevertheless, given the fact that Homer probably never traveled beyond his eastern Aegean world, his comprehension of the earth was widespread, including the Greek heartland, Anatolia, the Levant, Egypt and some of Africa to its west, the upper Nile, Sicily, and southern Italy, with some feeling of an encircling Ocean and perhaps its tides. As such, it is demonstrative of Greek knowledge of the eighth century BC. Other than the Catalogue of Ships there is little pure geography in the Homeric poems, but nevertheless an expanding sense of the inhabited world, and occasional comments about details of the earth's surface, such as the topography of the Troad, local winds, and sailing directions.[69]

The Phoenicians

The Phoenicians played an important role in the development of Greek geographical thought. The name "Phoenician" (mentioned several times by Homer) refers to the inhabitants of several cities on the Levantine coast from the Orontes River to the frontier of Egypt, most notably Byblos, Sidon, and Tyre, who shared a cultural identity but were not a unified state. They became prominent as early as the tenth century BC, most notably when the Israelite king, Solomon, commissioned Hiram of Tyre as the supplier for the building of the Jerusalem Temple, and to construct a fleet for him on the Red Sea.[70] Every three years Solomon also sent a Phoenician fleet to Tarshish, almost certainly the southwestern Iberian peninsula.[71] The Phoenicians became the great sailors of the Mediterranean, as Homer knew: he called them *nausiklytoi*, "famous for their ships."[72] They fought in many of the sea battles of the Mediterranean, most notably Salamis, and were constantly innovative in their ship building, probably inventing the bireme, the galley with two banks of oars.[73] Homer also knew of their commercial fame, although he saw them more as pirates than traders,[74] and from the end of the second millenium BC to Hellenistic times Phoenician ships roamed the Mediterranean and beyond. At an early date they attempted to establish settlements in the northern Aegean,[75] yet these evaporated as Greeks moved into the region in the seventh century BC. But the real Phoenician achievement was opening up the western Mediterranean, including the north African coast beyond Egypt, the western islands, and the southern Iberian peninsula.[76] Their earliest settlements were probably Utica (in modern Tunisia), Gadeira (modern Cádiz in Spain) and Lixos (on the Atlantic coast of Morocco), allegedly (but improbably) from the thirteenth century BC.[77]

Archaeological evidence actually suggests a date in the eighth century BC for the beginnings of western Phoenician settlement,[78] although the most famous Phoenician outpost, Carthage (Karchedon in Greek), was said to have been founded around 814 BC.[79] This implies a general date of the latter ninth and into the eighth centuries BC for Phoenician expansion into the west. At their peak, Phoenician towns and trading posts extended from Leptis in modern Libya west along the coast and out into the Atlantic, along the southern Iberian peninsula and the Lusitanian coast, and on to Malta, western Sicily, Sardinia, and the Balearics. There were also trading contacts in central Italy, as well as in the hinterlands of the regions that they settled.[80]

Around 600 BC, Phoenicians were commissioned by the Egyptian king, Necho (610–595 BC), to circumnavigate Africa.[81] Although this is one of the most famous stories from ancient exploration, it remains obscure. It is only documented by Herodotos, in an account written somewhat more than a century later.[82] The report is both tantalizing and brief, with none of the personal touches or names of individuals that characterizes Herodotos' writings. His source is Egyptian, as the Mediterranean is called the "northern sea." The context of the cruise is reasonable, since Necho was known for grandiose projects, such as a canal from the Red Sea to the Nile. Yet it is difficult to believe that at this time there was any conception of the size and shape of Africa, even though the Egyptians had long gone beyond the mouth of the Red Sea into the adjacent Indian Ocean.[83] This, coupled with the Phoenician activity in the Atlantic, may have given rise to the idea that the continent could be circumnavigated, especially if it were believed that Africa was long and narrow.[84] The Phoenicians might have felt that it was only a short distance from the mouth of the Red Sea to their cities on the Atlantic—this was thought to be the case as late as the end of the second century BC[85]—and a circumnavigation would have afforded easier access to these settlements.

Herodotos' sparse account provides few facts. The Phoenicians were to sail beyond the end of the Red Sea until they reached the Pillars of Herakles and entered the Mediterranean. The journey took more than two years, and each autumn they would land and plant a crop, waiting for it to mature before they moved on. This is perfectly reasonable, and Eudoxos of Kyzikos—probably familiar with this account—would do the same thing nearly 500 years later.[86] But the most interesting fact is that "in sailing around Libya they had the sun on their right," in other words, it was in the north. This shows that they were far south of the equator, perhaps even south of the Tropic of Capricorn, or close to the southern end of Africa. The sun was on the right (the account does not say "in the north"), which implies a westward journey of some distance, perhaps at the southern end of Africa where the coast runs east-west for more than 500 miles.

There seems little doubt that this astounding voyage of about 13,000 miles did take place, although this has not kept critics from ancient to modern times from rejecting it, beginning with Herodotos himself, who objected to the positioning of the sun, the very proof that the explorers were well into the southern hemisphere. In the early first century BC, Poseidonios of Apameia found the story equally untenable, largely because of its vagueness, and a century later Strabo felt the same way.[87] In fact the lack of specifics is a point in support, because there would be the irresistible temptation to elaborate a made-up tale with Homeric details about the strange peoples and navigational hazards encountered.

The exact details of the journey will never be known. It seems most likely that at the very least Necho commissioned a reconnaissance of the coastal regions beyond the mouth of the Red Sea, an area already partially known to the Egyptians. The story may even have become mixed with later reports of circumnavigation.[88] Nevertheless the time recorded for the distance traveled would allow the journey to be completed with relative ease even if half of it were in port.[89] However far Necho's Phoenicians went, and whether they did reach the settlements in west Africa, as seems possible, it was the last great effort of Phoenician exploration.

Egyptian and Babylonian pressures weakened the Phoenician homeland by the sixth century BC,[90] and the cities in the west became independent states, if this had not happened earlier. Eventually Carthage rose to a position of power, and established settlements of its own. But decline set in by the fourth century BC, and travelers reported that by this time there were deserted Carthaginian settlements along the Atlantic coast.[91] Nevertheless the Phoenicians and Carthaginians played an important role in Greek understanding of the western Mediterranean and the adjacent Atlantic. From the beginning, Greeks were informed about Phoenician activities, presumably through contacts in the Phoenician cities: such knowledge is implicit in Homer's data about the Ocean. Yet the Phoenicians left few written documents, and one is dependent on Greek and Roman sources for the extent of their travels, which have their own biases and misunderstanding of details. But Greeks were deeply interested in these matters, and as early as the fifth century BC Phoenician and Carthaginian travel reports were being summarized, or actually translated, into Greek.[92]

Early Theoretical Concepts of the Earth

The city of Miletos lies in southwestern Anatolia on the north side of the Maiandros River, whose twisting and turning inspired the word "meander." It was an ancient city with perhaps Cretan origins, a view supported by archaeology.[93] Later, the Milesians sent men to Troy, and in the generation after the war the city began to flourish, becoming the most important in the region.[94] The modern visitor to the site and its impressive remains finds that today the city is several miles from the sea, due to the inexorable siltation of the Maiandros, but during antiquity Miletos was a major seafaring city, located on a peninsula that jutted out into the river estuary. Eventually the Milesians founded dozens of settlements in the Black Sea and its approaches, probably beginning as early as the eighth century BC, although the dates of the first ones are uncertain.[95] By 600 BC, the Black Sea was encircled by

Milesian outposts, including, along its northern coasts, some of the most remote Greek cities. Milesians were well positioned to develop an early interest in geography.

Around 600 BC there was an astonishing eruption of intellectual curiosity in Miletos. Thales, a prominent local citizen, is credited with beginning this movement.[96] He was active in the latter seventh and the early sixth centuries BC and was the first Greek to consider the shape and positioning of the earth. As Aristotle reported:

> Others say that it [the earth] rests on water. This is the oldest concept that has come down to us (which they say Thales the Milesian believed), and it remained in position because it floated like wood or something similar.[97]

There is a clear analogy with a ship: the word "floated" (*ploten*) is a seafaring word. Seneca was more explicit, saying that Thales believed the earth was supported like a ship, and earthquakes were caused by the movement of the surrounding water.[98] It is certainly no accident that Thales should conceive of the earth through a nautical analogy—even if his views were directed more to an overall cosmology rather than geographical details—since he lived in a city particularly devoted to long sea voyages. Moreover, he also believed that water was the first principle of everything.[99] His theorization is primitive, and Aristotle sardonically asked what supported the water on which the earth floated. Nevertheless Thales was the earliest to conceive of the totality of the earth and to attempt to place it within the overall conception of the universe.

Anaximandros of Miletos—allegedly the student and successor of Thales[100]—refined his predecessor's thoughts by seeing the world as a stone column drum, with one of the flat surfaces the area occupied by humans. Rather than floating in water, the earth was supported by nothing.[101] This, too, very much reflected the environment in which Anaximandros lived: monumental stone architecture was becoming commonplace in the Greek world of the sixth century BC, and Anaximandros—who traveled at least from Miletos to

Sparta—would have been able to see the great temples of Archaic Greece under construction.[102] Anaximandros' ideas are more connected than those of Thales to what might be called nascent geographical thought: Eratosthenes of Kyrene, of the latter third century BC and the actual founder of the discipline of geography, called him one of the first geographers.[103] Most important in this respect was that Anaximandros was said to have been the first Greek to make a map. It is cited twice in later sources: Eratosthenes wrote of a "geographical plan," and Agathemeros (of uncertain Roman date), believed that it was "a plan of the inhabited world."[104] Moreover, Herodotos disdainfully referred to maps, but whether this included that of Anaximandros is by no means certain.[105] The word used by both Eratosthenes and Agathemeros for "plan"—*pinax*—originally meant a board or plank: to Homer it was the planks of a ship, a platter on which food was served, or, most interestingly, a writing tablet, cited as such only in the famous passage regarding the "deadly signs" given to Bellerophontes.[106] This last definition has a sense of graphic depiction that is close to a map, but the nature of Anaximandros' plan remains vague. It showed the inhabited world, and it is easy to think of it as having a political meaning, perhaps locating the many Milesian settlements. It was possibly similar to the famous map that Anaximandros' fellow citizen, Aristagoras, somewhat over half a century later, showed the Spartans in his futile attempt to enlist them in the Ionian Revolt.[107] Anaximandros also considered the nature of the sea and meteorology, thus moving closer to an overall attempt to understand the earth and its phenomena.[108]

Anaximenes, the third and last of the early Milesian natural philosophers, active around the middle of the sixth century BC, was also interested in meteorological phenomena, and believed that the earth floated in air and was flat, like a leaf.[109] Earthquakes were caused when the earth became wet and then dried out and cracked. He also thought that the sun was hidden by the higher parts of the earth at night, suggesting mountains in the far north.[110] Although it is problematic whether Anaximenes applied this theory to any practical suggestions of topography, the fact remains that this is an early attempt to theorize about details of the earth's surface beyond what was actually visible to those living in the Mediterranean.

The three Milesians, although cosmologists and natural philosophers rather than geographers, were the first Greeks to think about the nature and shape of the earth, its position in the universe, some of its specific phenomena, and, with the map of Anaximandros, the location of human beings on its surface. In addition there would have been practical observations from the Milesian seamen who already had covered much of the eastern Mediterranean and the Black Sea, although their data had not as yet found their way into theoretical speculation. Most notably, it was easy to see that the surface of the earth was curved, as ships and landforms sank beneath or rose above the horizon. The differing lengths of day and night at various latitudes, particularly as one went north to the remote Milesian settlements on the far side of the Black Sea, might also have been noticeable. More specifically, sailors going north would have seen changes in astronomical phenomena. The Bear had long been used to navigate, and it seems improbable that sailors did not notice that it rose higher in the sky as one went north. All this could provide suggestions about the shape of the earth and its position in the cosmos.

After Anaximenes, Miletos was no longer the center of cosmological speculation, which began to flourish in southern Italy and Sicily. Pythagoras of Samos, active in the latter sixth century BC, traveled widely and eventually ended up at Kroton in Italy, indicative of the broader world of late Archaic Greece that encouraged geographical thought. Pythagoras the individual remains a vague personality who is difficult to separate from his followers—both in his lifetime and afterward—and his own career has a mystical quality that is hard to analyze, but either he or his school may have made the most important early conclusion about the nature of the earth: that it was a sphere.[111] This revolutionary idea was also attributed to Parmenides of Elea, of the fifth century BC. Since Parmenides also spent his professional career in southern Italy and studied with a Pythagorean, Ameinias, it seems probable that he learned about the theory from the Pythagoreans, and may have been the first to publish it. But the exact origin of the concept remains obscure, and it is not definitely documented until the time of Plato.[112] Despite the widening of the Greek horizon, there is little evidence that the theory of a spherical earth was empirically based: it was merely a reflection of Pythagorean concepts of number, harmony, and perfection. It did not win quick acceptance: Sokrates was still concerned whether the earth was flat or round—meaning, presumably, a sphere—and as late as the third century BC, Eratosthenes had to remind his readers that his calculations presumed a spherical earth.[113]

Another inscrutable personality from Archaic Greece, Xenophanes of Kolophon, probably a younger contemporary of Pythagoras, was forced to leave his home and ended up in eastern Sicily.[114] He was a misanthrope who spent much of his time attacking his predecessors—from Homer to Pythagoras—yet made some interesting contributions to emergent geographical thought. Two of his poems—one on the founding of Kolophon, and the other on Elea—must have had some geographical context, probably more historical than theoretical, and they represent the emergence of a genre—city histories—that would have a strong geographical component. The latter work demonstrates some contact with Elea, the home of Parmenides (although Xenophanes belonged at least to the previous generation) and a closeness to south Italian philosophical thought. Xenophanes made some of the first comments about non-Greek ethnicity, pointing out the physical differences of Africans and perhaps Thracians;[115] consideration of ethnicity and its effects would become an essential part of geography. There is also the peculiar statement that the sun "failed" (*ekleipsen*) for an entire month, which seems to be impossible and unlikely to be a misunderstanding of an eclipse, but perhaps some uncertain knowledge of the long night of the far north.[116] Xenophanes also considered matters of changes in the surface of the earth, having found sea shells in the quarries of Syracuse and on Paros and Malta.[117]

Thus by the latter sixth century BC there had been scattered theorization about the shape of the earth, and its position in the cosmos, as well as rudimentary attempts to understand the changes that took place as one moved around the earth's surface, to some extent stimulated by the spread of sailors and scholars from the Greek homeland, especially into the Black Sea and Italy. To be sure, these speculations remained more a matter of cosmology and, as yet, there was little interest in landforms, natural processes, or demography, but they helped lay the basis for the development of a true discipline of geography.

CHAPTER 2

THE EXPANSION OF THE GREEK GEOGRAPHICAL HORIZON

Founding New Cities (Map 4)

Greeks were on the move and founding new cities throughout much of their history. But from the early eighth century to the latter sixth century BC there was a particularly intensive period of shifting population, as over a hundred settlements were established, from Tanais at the far northeast corner of the Black Sea to Emporion in northern Iberia, and Kyrene west of Egypt. In fact, much of the available coast of the Mediterranean came to be populated by Greeks during these years: they only tended to avoid areas already thickly inhabited (such as the Levant and central Italy), or regions that were claimed by others, especially the Phoenicians or Carthaginians in the western Mediterranean. Many of these new towns were at or near the mouths of the great European rivers: Massalia near the Rhodanos (Rhone), Spina near the Padus (Po), Istros and Tomis near the Istros (Danube), Olbia at the Borysthenes (Dnieper), and Tanais at the river of the same name (Don). All these rivers, and many others, provided access to a hinterland and trade routes that led from the Mediterranean or Black Sea far into the interior of Europe and to the North Atlantic and Baltic. The geographical impact of these settlements and their trading connections can be seen in the expansion of topographic knowledge. Homer's world was limited to the

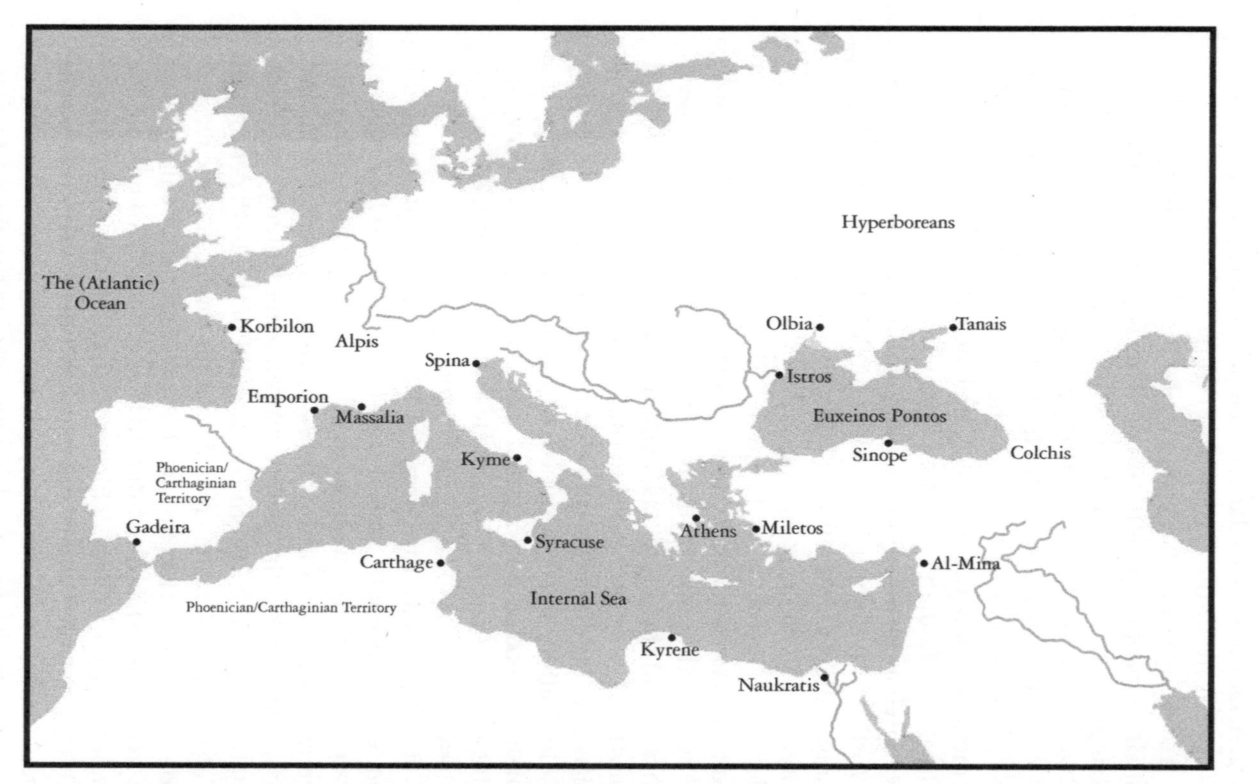

Map 4. The spread of Greek settlement in the Archaic period.

eastern Mediterranean and no farther west than Sicily. Hesiod, a generation or two later, knew about the Black Sea, the Etruscans in northern Italy, and Liguria.[1]

For many years it has been popular to call this dissemination of Greek peoples "colonization." This is an unfortunate term that grew out of false parallels with European activities of the sixteenth through nineteenth centuries. As early as the mid-sixteenth century, Spanish chroniclers were seeing parallels between Roman Spain and their own settlement in the Americas. Soon the analogies included the Greeks, and the Greek and Roman expansion throughout the Mediterranean was used as justification for the Spanish conquest of the New World.[2] Although there are some similarities, especially in the causes, there are significant differences, most notably that the Greek towns were generally independent states, not economic dependencies of a mother city.[3] The relationship between settlement and mother city was a matter of a shared ancestry, religion, other cultural institutions, and their dialect of Greek. Ties could be close, as demonstrated by the events in northwest Greece that led to the beginning of the Peloponnesian War, but the new cities also asserted their independence.[4] Colonization, as it was practiced in early modern times, is a highly imperfect analogy that seems anachronistic today, but the terminology is probably too deeply embedded ever to change, even in today's post-colonial world.

The reasons for this spread of Greek peoples are predictably varied. In antiquity, lack of land was believed to be a primary cause, and Plato saw it as due to overpopulation—likening emigration to a swarm of bees—as well as internal strife and external attacks.[5] Instability was endemic among the Greek states, and in the eastern Greek world of the sixth century BC Persian expansionism created another threat. Specific reasons for overseas settlement are known about particular states: Kyrene was established due to a seven-year drought in the mother city of Thera, which destroyed almost all the trees.[6] Often, someone who was at odds with his government would be encouraged to emigrate with his followers: Phalanthos of Sparta, after leading a failed coup, was sent to found Taras in

Italy.[7] The establishment of new cities could be tortuous and difficult, and there are many reports of failed settlements, sequential attempts to locate in several places, and even multiple returns to the mother city. The best surviving account of the process is the story of the establishment of Kyrene, around 630 BC, which took years to implement and was vividly described in great detail by Herodotos.[8] To be sure, one parallel with later European colonization remains valid: there was always a desire for wealth through trade (or less benign means), but in antiquity this was probably more a result of the new settlements, not a reason for them. Nevertheless, potential sources of wealth were identified at an early date, and when Kolaios of Samos sent 60 talents home from southwestern Iberia, about 630 BC, he had already heard something about the riches of little-known lands in the west.[9]

About 20 central and east Greek states were involved in these movements. The most prolific was Miletos, responsible for nearly a quarter of the settlements, which, starting around 700 BC, ringed the Black Sea and its approaches. This brought an extensive area into the Greek horizon, early enough for Hesiod to include two of its rivers (the Istros and Phasis) in the *Theogony*.[10] Kyzikos in the Propontis and Sinope on the south shore of the sea were probably the earliest; the Milesians were replicating the route followed by the Argonauts long previously.[11] The great rivers that flow into the northern Black Sea provided access to much of the interior of the eastern part of Europe, and relationships developed with the local potentates: Herodotos told the story of a certain Skyles, styled "king of the Skythians," who learned Greek and became Hellenized, alternating his residence between the Milesian city of Olbia and among his people in the hinterland, changing his dress appropriately.[12] Greek material goods penetrated far into the interior, up the Borysthenes (modern Dnieper) and other rivers.[13] On the west side of the Black Sea the Istros (modern Danube, which was the ancient name—as Danuvius—for its upper course),[14] extended nearly 1,800 miles into the interior, and the Milesian city of Istros was founded at its mouth, perhaps in the seventh century BC. Before long, dependent settlements were established a short distance up the river.[15] A rough topographical sense of the interior began to develop: it came to be believed that one of the Istros' upper tributaries was the Alpis, and that the sources of the great river were in the far west of Europe where the Kelts lived, somewhere around the city of Pyrene.[16] These earliest citations of the Alps and the Pyrenees, certainly known well before Herodotos recorded them, are rare documentation of the evolution of specific topographical knowledge, in this case frozen rather erroneously as early data.

Greek settlement in southern Italy and Sicily was more varied, but less valuable in the development of geographical knowledge, as Greeks had been in these regions since the Bronze Age. Several states from central Greece and the Peloponnesos established approximately 30 cities in this area. The earliest was by the Euboians on the island of Pithekoussai (modern Ischia) in the Bay of Naples, at the beginning of the eighth century BC, and before long—perhaps within a generation—Greeks had moved onto the nearby mainland at Kyme.[17] Greeks knew about the land to the north, but the Etruscan presence kept them out; nevertheless Greek trade goods soon penetrated their territory, representing the beginning of Greek contact with the central Italian world that later produced the Roman Republic.

There were no Greek settlements north of the Bay of Naples until the Ligurian coast (the modern French Riviera). On the eastern side of Italy there was a cluster of isolated towns or trading posts at the head of the Adriatic near the mouth of the Padus (modern Po), especially Spina (near modern Comacchio) and Adria (modern Atria). Greeks—the Athenians were perhaps the first—were here from the latter sixth century BC.[18] Spina and Adria may have been Etruscan outposts with a Greek mercantile quarter. Spina, however, was prosperous enough to endow a treasury at Delphi and for a while was the wealthiest city on the Adriatic.[19] These towns provided access to the Alps through the upper Padus valley, as well as to the amber route: Baltic amber had been known in the Greek world since prehistoric times.[20] The trade went through the passes of the eastern Alps and reached the Istros system, perhaps tapping the precious metal resources in modern Carinthia. But there never was a long-standing Greek presence at the head of the Adriatic, and any Greeks there were eventually assimilated by the expanding Etruscan and then Roman presence.

West of northern Italy was the Ligurian coast. On it was the most significant Greek city in the west, Massalia (modern Marseille), founded about 600 BC by Phokaia (modern Foça), an Ionian city that played a disproportionate role in opening the western Mediterranean to the Greeks.[21] Massalia lies at the head of a spacious harbor whose entrance is hidden, with freshwater springs emptying into it: a more propitious townsite can hardly be imagined, and its unusual topography is still apparent today. Moreover, this was the last good harbor before the mouth of the Rhodanos (modern Rhone), the largest river on this coast, no more than 30 miles to the west. The Rhodanos flowed directly from the north, and, ascending it, Massalian traders gained access to the river systems of northwest Europe. West of the site of modern Lyon the Liger (modern Loire) comes within fewer than 30 miles of the Rhodanos and, continuing up the Rhodanos and its tributary the Doubis (modern Doubs), it is eventually a crossing of only about 40 miles to the upper Rhenos (modern Rhine) at the site of Basel. Thus the Massalians had astonishingly easy access to much of northwest Europe, the North Sea, and the islands beyond. At the mouth of the Liger a trading post named Korbilon was established, perhaps at the site of modern Nantes.[22] Massalia founded other settlements, especially in the sixth century BC, when more immigrants arrived from Phokaia due to Persian pressures,[23] creating a sphere of influence that controlled the entire Ligurian coast, from Emporion (modern Empúrias in Spain) in the west to Monoikos (modern Monaco) in the east. Massalian trade and commerce still has its impact today: the first settlers brought vine cuttings with them that began the French wine industry.

There were no major Greek cities in Iberia beyond Emporion in the northeast, although there were a few trading posts:[24] Phokaians and Massalians did not venture to establish major settlements in Phoenician territory. But their explorers rounded the Iberian peninsula and may have gone as far as the British Isles, although there was little movement north of Ophioussa, perhaps Cape Roca near Lisbon, the westernmost point of Europe. Greek activity in and beyond the Phoenician zone was probably limited to a number of reconnaissances, and Korbilon, at the mouth of the Liger, was almost certainly settled from the interior. These explorations are described in a Latin poem of the fourth century AD called *Ora maritima*, by Rufus (or Rufius) Festus Avienus, describing the coast from Brittany to Massalia (in that direction), with some allusion to what was beyond Brittany. The poem is peculiar and difficult to analyze, as material over a wide chronological range is included—into the Roman period—but it seems in part based on a Phokaian *periplous* of the sixth or fifth century BC whose author is unknown. Confusing as the *Ora maritima* may be, it is the primary source for early Greek exploration of the coasts of the Iberian peninsula and beyond.[25]

North Africa was largely devoid of Greek settlement, except for those in the Kyrenaika west of Egypt, where Kyrene was founded by the Aegean island of Thera around 630 BC.[26] As it and its dependencies were the only Greek cities in a vast area, Kyrene served as an important contact point between the Greek world and interior Africa. By the mid-fifth century BC the oasis routes across the Sahara were known, with one reaching as far as a city and an eastward-flowing river containing crocodiles.[27] There is no evidence for the name of the city—Timbuktu and others have been suggested—but the river is probably the Niger.[28] Yet Herodotos' informants thought that it was the Nile, perhaps for no other reason than the Nile was the only major river known in Africa. The southernmost stages of the journey were guided by people who spoke no known language and were "small men, less in height than normal." This was not the only time that Herodotos referred to men of small stature in sub-Saharan Africa, since a Persian, Sataspes, around 479–465 BC, encountered them in a failed circumnavigation.[29] These were probably the earliest specific references to the pygmies of central Africa: they were known to Homer[30] and appear on the François Vase of the early sixth century BC, but in these instances were generic rather than specific. By contrast the "small men" in West Africa seem actually to have been encountered, although perhaps in a region somewhat north of their normal range, as least as it is understood today.[31]

In addition to these many settlements, Greeks established a handful of trading posts in areas that were already populated, most notably Naukratis in Egypt, about 40 miles up the Kanobic mouth of the Nile on the left bank. It was a joint foundation of at least 10 Greek cities, all in eastern Greece or the adjacent islands, with Miletos perhaps taking the lead, and at first holding a monopoly on Greek trade with Egypt. Herodotos implied that it was created at the urging of the Egyptian king, Amasis (570–526 BC), which is when it flourished, but pottery suggests an original date in the seventh century BC.[32] Needless to say it was different from a Greek town in a previously unsettled area, and the involvement of so many Greek states created a pan-Hellenic atmosphere unusual for the era, allowing Greeks to learn about Egyptian culture and civilization as well as what lay up the Nile and down the Red Sea. The tradition that Greek intellectuals visited Egypt as part of their education, beginning with Thales,[33] however debatable, is perhaps a metaphor for these cultural ties, and there is no doubt that prominent Greeks of the Archaic period spent time in Egypt or Naukratis.

In the Levant, there was a Greek post at al-Mina at the mouth of the Orontes River in Syria, just downstream from the site of the future city of Antioch-on-the-Orontes. Mycenaeans had been in the region and, from the ninth century BC, there were settlers, probably from Euboia. The flourishing period of the settlement was during the seventh and sixth centuries BC.[34] Because the evidence is archaeological rather than literary, only the material culture is known, but presumably al-Mina provided a Greek window onto both the Phoenician cities just to its south and interior Syria.

The importance of this expansion of the Greek world from the eighth through the sixth centuries BC cannot be overstated. By the time the era came to an end and the Greek world was turning its attention to another type of expansionism, that of the Persians, the Greek horizon extended far beyond the coast of the Mediterranean. Greeks living in distant cities such as Massalia and Olbia were learning about the interior of Europe; those in Kyrene and the eastern trading posts became aware of sub-Saharan Africa and interior Asia. The Mediterranean was no longer the limit of Greek geographical knowledge. In most cases the information that reached Greeks on the coast was derivative, and in fact few traveled far inland from their own cities. One who did was Aristeas, from Prokonnesos, near the entrance to the Black Sea, who penetrated into the regions far to the north of the sea. Herodotos dated him to the early seventh century BC but he may be somewhat later.[35] A figure hidden in myth, with cultic overtones, he also appeared more than 200 years later at Metapontion in southern Italy. He is credited with being the author of a hexameter poem called the *Arimaspeia* (named after one of the peoples he was seeking), which seems to have been still available in the first century BC,[36] yet both Aristeas and the poem are shadowy, and some of its extant fragments may be forgeries from the fourth or third centuries BC. The poem records a six-year journey north of the Black Sea, the first Greek penetration into this region.[37] One of the goals of Aristeas' trip was to find the even more mysterious Hyperboreans, those "beyond the north"—a descriptive term rather than an ethnym[38]—but he only went as far as the Issedonians, who nonetheless were a great distance to the east.[39] It has been intriguingly speculated that the Hyperboreans might have been the Chinese, although there is no proof of this and, even if so, there was no contact between them and the Mediterranean world for centuries.[40] It may be impossible to untangle the three personas of Aristeas: the legendary cultic figure, the traveler to the far north, and the author of the *Arimaspeia*, yet if the poem is genuinely from the seventh century BC, it is the first quasi-geographical work in Greek literature. Aristeas may even have been sent by the city of Prokonnesos in the early years of the settlement to reconnoiter the far north: such travels are reasonable as Greeks settled on the Black Sea and began to show interest in the potential value of its hinterland.

Yet it is in the nature of trade routes that rarely would any single person travel their entire distance (those who went from Kyrene to sub-Saharan Africa are an exception), but goods were handed off and received at various way stations to or from those traveling the next segment, a process repeated many times. Information on what lay beyond was transmitted the same way, but was subject to misinterpretation, or provided in such a manner as to be incomprehensible by the time it reached the coast: the misunderstanding of the Alps and Pyrenees is a good example. Nevertheless a large amount of information was collected during these centuries, and by the latter sixth century BC the Greek horizon was astonishingly broad, so that Hekataios of Miletos was able to report on essentially the entire Mediterranean and Black Sea coasts (with some exceptions for the Phoenician territory).

The First Greeks Outside of the Mediterranean

As early as the seventh century BC, Greeks began to sail beyond the Mediterranean and the Black Sea. There was only one way of leaving them: through the Pillars of Herakles. Everywhere else the seas were enclosed and their coasts had been explored. About 630 BC (a synchronism is provided with the foundation of Kyrene), a few generations before the Massalians penetrated the Atlantic, a ship captain named Kolaios, from Samos, who normally made the Samos–Egypt run, was blown off course to Platea, an island several hundred miles west of Egypt.[41] This diversion is not improbable, since if winds had prevented him from making Egypt, he might have headed for Platea, the only known Greek settlement on the north African coast, a predecessor of the future city of Kyrene. Putting out to sea from Platea, Kolaios allegedly headed for Egypt, a coastal sail of about 500 miles. Instead, an easterly wind blew him to the west, and he could not stop until he was through the Pillars of Herakles, a distance of about 1,100 miles. Fortuitously he ended up at Tartessos, biblical Tarshish, the wealthy region of the southwestern Iberian peninsula. He was said to be the first Greek to visit this area. He returned home to Samos with 60 talents of goods; the government of the island took a commission of 10 percent and used it to provide art for their famous temple of Hera.

This is all that is known about Kolaios, a story told only by Herodotos. There is little doubt about the results of the endeavor, and that it was the earliest Greek attempt to gain access to the wealth of southwestern Iberia. It is unlikely that Kolaios would have been welcomed in a Phoenician port such as Gadeira, especially if his motive was to tap into the area's resources for profit, so presumably he made some contact with the locals outside the Phoenician trade system and returned to Greece before he ran into difficulties. There is no reason to suspect any part of Kolaios' story except the storm of Homeric proportions that sent him to Tartessos. He may have decided, after ending up in Platea—outside the region where he was known—to strike out on his own and to act on the rumors of western wealth that had been penetrating back to Greece, eventually explaining the incident to his associates on Samos by means of the storm. Yet to be blown from Platea to Tartessos is essentially impossible: to say nothing of the matter of provisions, Kolaios would have had to negotiate the Sicilian Strait between Sicily and Africa, requiring changes of course. It was also remarkably convenient that the winds allowed him to thread the eye of the needle that was the narrow strait between the Pillars of Herakles and which led to the Atlantic. Kolaios belongs in that long tradition of seamen who claim to have been blown off course only to make a momentous discovery, but who probably knew all along where they were going and why. Nevertheless he made the world beyond the Mediterranean available to the Greeks. Shortly after his time, the Tartessian king, Arganthonios, who reigned 80 years, entered into a relationship with the Phokaians, perhaps an indirect result of Kolaios' voyage, and financed their defense against the Persians in the mid-sixth century BC.[42]

There is another hint of an early Greek traveler into the Atlantic, a certain Midakritos, who was the first to import *plumbum album* (tin) from Kassiteris Island. A single sentence in the *Natural History* of Pliny is the sole source for his activities.[43] "Kassiteris" is a Hellenized Elamite word meaning tin, known since earliest times,[44] so Midakritos went to the Tin Island and brought the metal back to the Mediterranean. No date is provided, but he must have been not much later than Kolaios. He may have been either Phokaian or Massalian. Tin originally came to the Greek world from the east, as the Elamite word demonstrates, but Midakritos represents the opening of new western supplies. The Kassiteris Island—the name is usually plural, Kassiterides—is not easily located: Herodotos suggested that they were in the far west but knew nothing about them.[45] The most detailed account of the islands is by Strabo, written hundreds of years after Midakritos and showing no awareness of him.[46] It relates how they were located in the open sea at the latitude of Prettanike (the British Isles) and opposite the European mainland, yet beyond Lusitania, somewhere up the Atlantic coast. There were ten of them, and they were inhabited by people in black cloaks who carried wands. The islands had mines of tin and lead, which were near the surface, originally exploited by Phoenicians from Gadeira, who kept their location hidden. The Romans unsuccessfully tried to find them, perhaps in the context of Caesar's activities in Gaul, but eventually they were forgotten. Modern topographical analysis has been inconclusive, and it is not even certain whether they were islands or coastal promontories, and no identifiable group of ten islands exists off the coast of Brittany or the British Isles, although the topography of this region has changed significantly. In early modern times the Scillies were the most popular suggestion, but they have fallen out of favor recently, and topographers probably no longer have any chance of locating them.[47]

Exact identification seems unnecessary to understand Midakritos' journey. In his quest for tin, he was remembered long thereafter—even if certain details of his voyage had been lost—and he went much farther than Kolaios—as far as northwestern France. Tin had been known in this region since the Late Bronze Age,[48] but any supply to the Greek world may have been disrupted by the Phoenicians, and Midakritos found a way to avoid them, much as Kolaios had.

It is possible that a report on the voyage of Midakritos was one of the sources buried in the *Ora maritima* of Avienus of the fourth century AD. This is purely speculation, but at the very least the anonymous *periplous* of the *Ora maritima* is from the same environment as Midakritos. The account begins at the unknown place known as the Oistrymnic Bay, a toponym associated with northwestern Brittany. It then moves south, passing coastal places that have not been located, until it reaches Tartessos and Gadeira, then the Pillars of Herakles and the Iberian and Ligurian coast, ending at Massalia. A considerable amount of sailing data is included, and the earlier and more remote parts are, expectedly, more uncertain than the description within the Mediterranean. Its extremely late date, and the use of the Latin language for material originally presented in Greek, create numerous problems of interpretation, but there is little doubt that it is evidence for an early exploration of the Atlantic coast, almost certainly Phokaian or Massalian, north of the Pillars and as far as Brittany. Mention of the Hierni and Albiones suggests knowledge of Ireland and the British Isles.

Midakritos' attempt to open up a sea route to western tin supplies was eventually futile. The rise of the Carthaginians in the latter sixth century BC disrupted these activities, and the voyages of Kolaios and Midakritos became isolated incidents rather than the beginnings of a continuous pattern of trade, although some imprecise details of their travels were remembered. By Roman times even the location of the Tin Islands had been lost.

The Massalians were also curious about what lay south of the Pillars. Learning about it was assigned to Euthymenes, who is little known today.[49] He is not cited by name in extant literature before the Roman period, when it was reported that he had sailed on the Atlantic Ocean. He also appears on a list of the first century AD positioned between Thales and

Anaxagoras, which would seem to place him around 500 BC. This was the time of the peak of Carthaginian exploration, and the Massalians may have heard enough to want to make their own reconnaissance. Euthymenes reached a west African river which was so large that fresh water went far out to sea, and which contained crocodiles and hippopotami.[50] He believed—in part because of their presence—that this river connected to the Nile, and he may have seen a strong flood tide that pushed water upstream. No other details are preserved.

It is not possible to identify the river with precision, but the most probable is the Senegal, the northernmost major tropical river of west Africa, which has the phenomena described. If Euthymenes went that far, it was an important journey, perhaps designed to inspect the Carthaginian presence, especially if the Massalians had learned about an explorer from Carthage, Hanno, who penetrated the same region at about the same time.[51] But Euthymenes seems to have learned nothing that interested the Massalians, who made no further effort to visit the west African coast. Like the voyage of Midakritos to the north, that of Euthymenes was largely forgotten and Greeks tended to stay off the Atlantic as long as the Carthaginians were in power.

The Effect of the Persians

The Greeks, seafarers by nature, usually did not explore the interior of the lands adjacent to the Mediterranean but relied instead on trading reports reaching the coast. But, at the eastern end of the Greek world, changes were occurring that had a profound effect on Greek geographical knowledge. Ever since the early sixth century BC the Anatolian Greeks had been subject to the political authority of the Lydian kingdom, centered three days inland at Sardis. Early in the following century, King Alyattes made a treaty with the eastern Greeks, who became highly prosperous through trade with Lydia. Alyattes' son, Kroisos (Croesus), who came to the throne about 560 BC and was fabled for his wealth, ruled with a somewhat heavier hand, yet the Greeks continued to flourish. Stories of

Greek advisors at Kroisos' court, most notably Thales of Miletos, whether or not true, are demonstrative of the symbiotic relationship between the eastern Greeks and Lydia. Then, suddenly, probably in 546 BC, Kroisos overreached himself and the Lydian kingdom was destroyed by Cyrus of Persia. The Persians—about whom the Greeks had previously heard little—inherited the Lydian kingdom, including its Greek possessions, inaugurating two centuries of a contentious relationship.[52] Whatever the political situation, Persian control of eastern Greece had its impact on geography, making Greeks aware of the great expanses of the inhabited world to their east. The Royal Road from western Anatolia to the heart of Persia became a major feature in Greek understanding of eastern geography.[53] First mentioned in Greek literature by Herodotos in the context of the visit of Aristagoras of Miletos to Kleomenes of Sparta in 500 BC,[54] it was already well known to the Ionian Greeks, with parts of the route probably existing since prehistoric times. Herodotos outlined its 111 *stathmoi* (stations or stages), measured in Persian parasangs, presumably obtained from an official Persian record available in Miletos or Sardis. The road continued to be of geographical significance into the Roman period. In the first century BC, Isidoros of Charax described portions of it in his account of the Parthian route from the Euphrates to Arachosia.[55]

The Persians, then, opened up the eastern horizon for the Greeks. Greeks also learned about Cyrus' homeland of Persia, and Cyrus himself ranged far, capturing Babylonia in 539 BC.[56] Much of the following decade was spent in an eastern expedition that ended up in the land of the Massagetians, east of the Caspian Sea, who killed Cyrus in the summer of 530 BC.[57] The circumstances of Cyrus' fate were widely reported, providing Greeks with their first knowledge of a world that extended to, and beyond, the Caspian.

Cyrus' son and successor, Kambyses, captured Egypt for the Persians in 525 BC, which put Naukratis under Persian control, yet he encouraged trade and many Greeks came to Egypt at this time.[58] He also allegedly went up the Nile to the First Cataract and sent a reconnaissance to investigate the Aithiopians, with a view to annexing them. A few of their northernmost regions were conquered but the value of the expedition was its report on their ethnography, although largely of a fantastic nature.[59] Yet Greeks became aware of exotic flora and fauna, such as ebony and elephants.

Kambyses' eventual successor, Dareios I, who came to the throne in 522 BC, sent an expedition far east to investigate India, a region unknown to the Greeks. The toponym "India" ("Indike" in Greek) originally referred merely to the Indus valley, although by Hellenistic times the term had expanded to include the entire sub-continent. The Persians had previously reached its borders and had subdued some of the Gandarans.[60] Dareios' explorers, who were in the region shortly before his Skythian expedition of 513 BC, did not go beyond the Indus valley, and even in the fifth century BC it was believed that everything to its east was a deserted sandy region.[61] The king was particularly interested in the Indus River, and commissioned an expedition to sail down the river and to return west to Persis. It started from Kaspatyros—whose location is unknown but which must be on the upper Indus system—and after 30 months came to the mouth of the Red Sea.[62] A Greek named Skylax, from Karyanda in Karia, took part in the cruise, and was asked to write a report, perhaps titled *Circuit of the Earth*, which included comments about flora and fauna, the social structure of the Indians (with the earliest allusion to the caste system), and other ethnographical data, including descriptions of anatomically improbable peoples.[63] Although the title of his treatise seems grandiose, Skylax's lengthy sail along the coast of the Indian Ocean, previously unknown to Greeks, may have made it seem that he had made a circuit of a significant part of the entire earth.

The scant fragments that survive of Skylax's treatise are confused: Herodotos said that the Indus headed east (the river actually runs south by southwest), and 30 months is a long time for the journey, but perhaps there was no reason for speed if it were an intelligence-gathering project. There are no surviving details of the actual cruise from the mouth of the

Indus to the Red Sea, and in fact these coasts remained essentially unknown until the time of Alexander the Great. Skylax's report stands in isolation: if it were a personal document for Dareios,[64] it may not have circulated widely and even Herodotos, slightly less than a century later, probably heard about it only through hearsay. Neither Eratosthenes, Poseidonios, or Strabo—the three most assiduous Greek writers on geography—seems to have known about the report, although Strabo knew about Skylax but provided no details.[65] Nevertheless Skylax's treatise, however ephemeral, is of great significance, for it is the first known Greek work to devote itself solely to a geographical topic. With Skylax the concept of geography and geographical writing were closer to coming into existence.[66] Moreover, the voyage of Skylax made the Persians—and thus, to some extent, the Greeks—aware of the Indian Ocean: Dareios I was said to have "made use of the sea," although the evidence of pre-Hellenistic travel on it is scant.[67]

Dareios was personally responsible for another expedition to the perimeters of the Persian world: against the Skythians in Europe, with "Skythian" the generic term for the peoples beyond the Black Sea. A contingent of Greeks accompanied the expedition, which set forth in 513 BC, bridging the Bosporos and then the Istros, and penetrating a short distance beyond the latter. This was territory known to the Greeks since the time of the first settlements along the Black Sea (hence their involvement),[68] but the journey provided an opportunity to compile a detailed ethnography of the Skythians and their unusual habits,[69] firmly placing the customs of remote peoples within the emergent geographical tradition.

Thus by the end of the sixth century BC the circuit of Greek geographical knowledge included the Atlantic coast of Europe as far as the British Isles and perhaps Ireland (although these were little more than toponyms of uncertain location). The rivers of western Europe were known, but the Alps and anything that lay to the north of them were still vague, except perhaps for the amber route to the Baltic. The coast of the Black Sea was heavily populated, and there was some understanding of the adjacent interior and the peoples who lived along the great rivers, but with many uncertainties and gaps. Most of Anatolia had been explored, and the Persians provided selective data about Babylonia and Persis, as well as some of the Caspian region and India, and the routes to those places. Yet entire regions within this perimeter, such as the eastern Iranian plateau, were barely comprehended. Skylax had seen the coast of Asia and Arabia but seems to have provided little detail about it. The Red Sea, upper Nile, parts of Aithiopia, and the east African coast were known from Egyptian and Persian sources, and it was understood that Africa could be circumnavigated, but there was no information about the interior of the continent beyond a few caravan routes leading south across the Sahara. The Phoenician settlements on the Atlantic coast provided some awareness of that region. Surrounding all these lands was the great External Ocean, comprehended in some uncertain way since the time of Homer, and whose existence was no longer doubted. It connected to all other seas, including the Mediterranean, Black, and Red Seas, with the only exception the Caspian, which was an enclosed sea.[70] Data were also being gathered about those who lived in these regions and on these coasts, and what their customs were, yet it is doubtful that this information was coordinated in any particular way. The exact relationship of the landmass to the External Ocean was as yet undefined, and no one had seen any part of the Ocean from the Atlantic coast of France around to the north and east, and on to the west coast of India. Yet the leading minds of the era were coming to the conclusion that the earth was a sphere.

Hekataios of Miletos

It was Hekataios of Miletos who wrote the first general geographical treatise.[71] A distinguished citizen of his home city, he (futilely) advised Milesians not to revolt from Persia in 500 BC, using his geographical and historical expertise to suggest that engaging the Persians was unwise because of their great extent and power, perhaps the first documented example of the use of geography for political reasons. He also said that the Milesians' only chance for survival was to develop their sea power. When the revolt and subsequent Persian retaliation came, Hekataios was one of the peace negotiators. He spent time in Egypt: unlike the formulaic association with Egypt on the part of earlier Milesian intellectuals, there is enough detail not to doubt the story. He was also said to have traveled widely, and came to be seen as a pioneer of geographical research.[72]

Hekataios' geographical treatise was either titled *Periegesis* (*Leading Around*) or *Periodos Ges* (*Circuit of the Earth*).[73] He was an innovative author, one of the first to write in prose, and his titles represent new terminology, although Skylax may have used the latter one a few years previously. Herodotos, for whom Hekataios was an important source, called him a *logopoios*, a "maker of stories," not a pejorative term but one that also relates to the use of prose.[74]

Although the treatise was still available in Hellenistic times—Eratosthenes and Strabo consulted it—its authenticity became disputed, and, as geographical knowledge increased, it became less viable.[75] Today there are more than 300 fragments preserved, most of which are simple topographical entries in the late antique *Ethnika* of Stephanos of Byzantion, probably from the sixth century AD. Many of them consist merely of the toponym and a note as to whether it is from the European or Asian book of the treatise, demonstrating the wide range of the work but adding no local color. The division of the toponyms into these two categories is probably the earliest attempt to understand the inhabited world by means of continents, although the word "continent" (*epeiros*) does not appear in the extant fragments, which proves nothing: it merely meant "land" to Homer (what the Achaians drew their ships onto), and its first extant use as "continent" was by Aeschylus, as Atossa grimly remarked that Xerxes depopulated an entire continent.[76] To Hekataios there were only two continents, with their division uncertain. Herodotos knew two possible lines of separation: either Lake Maiotis and the Tanais River (the modern Sea of Azov and Don River) or the Phasis (modern Rioni) River.[77] The former prevailed, accepted by the fifth-century BC author of *Airs, Waters, and Places.*[78] Hekataios seems to have had no conception of Libya (Africa) as a third continent, since fragments from that part of the world are variously ascribed to Egypt, Asia, Libya, or merely to the *Periegesis.*[79] By the time of Herodotos the third continent of Libya was taken for granted, and there were no new continents until the discovery of the New World, although the Pythagoreans suggested the existence of others. Whether or not this was so, it became a persistent topic from the fourth century BC.[80]

Any theory that Hekataios may have had about continents cannot be determined from the extant fragments. Eratosthenes told how Greeks first came to Karia in Anatolia, saw that things were different, and that somehow continental theory developed from this encounter.[81] But he was uncertain about the tale, which sounds more anecdotal than scientific, and Strabo, the extant source, chided Eratosthenes for not expressing himself clearly. Nevertheless the story does imply that differences in ethnicity may have been behind the first ideas about continents.

Hekataios may have made a map, although the evidence is uncertain and is subject to the usual problems of the Greek terminology for cartography.[82] It is said that he improved the map of Anaximandros, but in a famous passage Herodotos ridiculed mapmakers who made the world symmetrically round, with Asia and Europe of equal size.[83] The citation of only two continents—when Herodotos knew full well that there were three—suggests Hekataios as a source, although Herodotos' polemic includes more than one map maker. Maps were certainly topical in the Miletos of Hekataios, where Aristagoras used one (perhaps even drawn by Hekataios) as a political tool, with unfortunate results.[84]

Possibly connected with an interest in maps was the development of the concept of the meridian: the north–south line connecting points of equal longitude. The Greek word for meridian, *mesembria,* means "midday," since when it is noon at a given point, it is also noon at all points on the same meridian. The word (in its geographical sense) is not documented before the fourth century BC, but the concept is earlier, and may have originated with Hekataios.[85] The first extant attempt to create a meridian is Herodotos' statement that Egypt, Kilikia, Sinope, and the mouth of the Istros, lay on the same line, exceedingly rough but creating a theoretical connection of far-distant points.[86]

Hekataios' circuit of the earth began in the west of Europe, perhaps using data from Phokaia.[87] The extant fragments for this region—totally from Stephanos—are names of places and peoples, many of which do not appear in later sources and cannot be identified, but include Tartessos (as a region). The six names from Liguria seem to be limited to the coastal regions, and there is no apparent knowlege of the interior. The topography of Italy includes only the area from the Bay of Naples to the south, except for the islands of Aithalia and Kyrnos (modern Elba and Corsica), with nearly half the names coming from Sicily, representing a world that knew little of the Etruscans (only Aithalia was said to be Etruscan), and where the name "Italy" was still limited to the far south, its place of origin.[88]

In a rare preserved ethnographical statement among Stephanos' toponyms, Hekataios wrote about the prosperity of the Adriatic coast, where the cattle produced calves twice a year and the hens laid eggs twice a day, perhaps an allusion to the fertile lands settled by the Greeks at the head of the Adriatic.[89] He knew about the Istrian peninsula and the Illyrian coast, as well as the river system of Epeiros;[90] Strabo's extensive use of the earlier scholar for this region demonstrates that the information was still valuable 500 years later. In addition to the usual toponyms preserved by Stephanos, material on Greece proper includes a long passage on the Pelasgians—the pre-Greek peoples of the Greek peninsula—and their relationship to the Athenians, paraphrased by Herodotos, although this may be from an historical work by Hekataios, not his geographical one.[91] Whatever the source, it shows an early interest in Greek origins and city history, which looks ahead to the efforts of Herodotos slightly later, and demonstrates that history and cultural geography had become intertwined. Other thoughts about the early demographic history of the Greek world—in particular the Peloponnesos—were preserved by Strabo.[92]

There are fragments about the Greek islands, Macedonia, and Thrace, but these are almost all isolated toponyms, and the southern Greek islands and Crete are not among the preserved citations.[93] Ethnographic comments are rare, except for the peculiar drinking habits of the Paionians, noting that they made oil from milk, perhaps the earliest reference to butter, which was not generally part of the Greek diet.[94] Data on Skythia are limited, but it is possible that Herodotos' extensive discussion of the Skythians relied on Hekataios.[95] There are two references to peoples in the Caucasus, which is defined as being in Europe, but it is not clear where the division between the two books of the *Periegesis* comes, and it is probable that the border between the continents had not been precisely established and there was an overlap.[96]

There are a number of citations of places on the Asian coast of the Black Sea and in the Troad, mostly familiar names, as well as the hint that Hekataios may have included some Homeric topographical exegesis (Strabo, an expert on the topic, was not convinced by his arguments).[97] Many fragments survive from the remainder of Anatolia, as one might expect, but with little detail.[98] The Levant and interior Asia are poorly represented, yet the account extended as far as India, which was perhaps some of the most recent material in the treatise, implying that the report of Skylax had penetrated to Miletos.[99] Hekataios also knew about the Caspian Sea and its topography and flora.[100] He seemed uncertain about the data and used the alternative name "Hyrkanian" for the sea, which may have originally been a local ethnym. This was again recent information, probably from the expedition of Cyrus the Great in the 530s BC.

Hekataios' description of Egypt was a prime source of material for later writers, and authors such as Herodotos, Diodoros, and Arrian were indebted to him.[101] Hekataios spent time there, and was probably the first Greek to view Egyptian customs with a critical scholarly eye. Not much later

Herodotos seems to have had Hekataios' treatise in hand as he traveled around Egypt, and the slightly polemic tone that characterizes his Book 2 is probably due to an urge to refute Hekataios, who is mentioned by name only once, on a matter of genealogy.[102] Despite the reliance of later authors on Hekataios' description of Egypt, little identifiable detail has been preserved.

The rest of Africa—as Hekataios knew it—is contained in the remaining fragments. There are only four about the Aithiopians and their world, but again he may have provided data for Herodotos, having used the report of Kambyses from the 520s BC.[103] He did discuss the military tactics of the pygmies, and since it is known that he indulged in Homeric criticism, this may have been an exegesis of the famous comment on pygmies and cranes at *Iliad* 3.6. There are more fragments about Africa west of Egypt (ancient Libya) but the terminology is inconsistent and there is no explicit identification of Libya as a third continent. A number of Carthaginian settlements are mentioned, the earliest reference to this great power.

The frustrating nature of the fragments of Hekataios' geographical treatise does not hide its importance. His access to several hundred toponyms throughout the Mediterranean and Black Sea demonstrates the advance of Greek geographical knowledge in the previous centuries. In addition to the place names that overwhelm the modern survival of the work, there was also mythology and ethnology, and even Homeric topographical criticism. Although specifics are difficult to come by, his knowledge of Skylax, Kambyses in Egypt, and Dareios in Skythia implies composition after 515 BC. Except for interior Asia and up the Nile, the data are coastal: there is little if any comprehension of the interior of Europe. Lacking also is any political history, for despite Hekataios' role as a statesman, there is no hint of the Persians—even though Miletos was Persian territory in his day—or of the Carthaginians or Etruscans, except that a few cities were said to belong to them. He probably did not see contemporary politics as suitable for a geographical work, perhaps reserving such matters for his more historical treatise, the *Genealogiai*.[104] The opening of this work deserves particular note as the most explicit extant statement about Hekataios' theory of scholarship:

> Hekataios the Milesian says the following: I write these things since they seem to me to be true, for the stories of the Hellenes—as they appear to me—are numerous and laughable.[105]

This statement was well known to Herodotos[106] and firmly places Hekataios in the new world of scholarly ethnography and geography, not myth and fantasy. Hundreds of years later Strabo put him at the forefront of early geographical scholarship—alongside Homer and Anaximandros—and further honored Hekataios by following his pattern of a clockwise circuit of the inhabited world.[107]

CHAPTER 3

THE SPREAD OF GEOGRAPHICAL KNOWLEDGE AND SCHOLARSHIP IN THE CLASSICAL PERIOD

The Carthaginians

It is not known for certain when Carthage became independent from Phoenician control—perhaps as early as the eighth century BC. When this occurred, the Carthaginians acquired the Phoenician possessions in the western Mediterranean and on the Atlantic coast. They also embarked on their own program of exploration, seeking trading and mercantile opportunities. Around 500 BC, Himilko and Hanno, two members of the ruling Magonid family, were sent to explore the region beyond the Pillars of Herakles. A summary report of Hanno's journey survives in a Greek translation probably of the fifth century BC, the earliest extant *periplous*.[1] In Greek sources, Hanno was first cited by name in the Aristotelian *On Marvellous Things Heard*, although Herodotos probably knew about him. Despite intense modern criticism, there is no reason to doubt that the voyage actually took place.[2]

The text that exists today is a summary—whether the work of Hanno or the Greek translator is unclear—with several gaps and a sudden conclusion that gives the impression that the translator lost interest. The account has been partially Hellenized, with some toponyms translated into Greek descriptive terms, and using "Aithiopian" for one of the peoples of sub-Saharan Africa, a purely Greek view of the demographics of this region. The text opens with a statement of purpose, which suggests a major colonizing expedition:

> It was decreed by the Carthaginians that Hanno sail beyond the Pillars of Herakles and establish Libyphoenician cities. Thus he sailed in command of 60 fifty-oared ships and with a great number of men and women, in the amount of 30,000, along with grain and other supplies.[3]

Any account of the journey from Carthage to the Pillars has not been preserved, and the more detailed narrative begins when they reached Thymiaterion, two days beyond the Pillars. They continued to Soloeis, probably around Cape Spartel, the northwest point of Africa, and headed along the Atlantic coast, establishing settlements (perhaps small trading posts) as they went. Eventually they reached the Lixos River, where they stayed for a while, possibly making a reconnaissance into the interior, and taking interpreters on board, which would suggest that the Carthaginians had already been this far. The town of Lixos at the mouth of the river (modern Leukos in Morocco, which preserves the name) is the one place that Hanno visited that continued to be important into Greek and Roman times, although there are insoluble difficulties in coordinating Hanno's toponym with the historical site. Farther down the coast they also had an extended layover at a place they named Kerne, which remained an Carthaginian outpost for some time, but which cannot be located today.[4]

After Kerne the character of the account changes. There is no further indication of establishing settlements or extended stops, and one has the impression that only Hanno and a few companions went farther. The land became remote and mysterious, with increasingly hostile locals and little help from the interpreters:

> we landed, seeing nothing in the daytime but woods, yet at night many fires were burning, and we heard the sound

> of flutes, cymbals, and the beating of drums, and an infinite amount of shouting. We were taken with fear, and the seers ordered us to leave the island. We quickly sailed away and passed by a land that was full of burning incense, from which fiery streams flowed down to the ocean.[5]

Strange flora and fauna were encountered, including crocodiles and hippopotami. Volcanic phenomena were also prevalent. The expedition turned east at the Horn of the West, probably the Île de Gorée in Senegal, and eventually reached the Chariot of the Gods, a huge mountain that put forth fire. There is little doubt that this is Mt Cameroon, the only active volcano on the coast. Three days later, about a month from Carthage, they reached the Horn of the South, probably modern Cape Lopez in Gabon, the westernmost point on the southern African coast. Here they encountered a wild and hostile people called the Gorillai, a local ethnym that in the nineteenth century was misapplied to the species of ape. Then, suddenly, the expedition turned back, although there may be a gap in the extant text at this point. They were far from home (at about the equator), low on supplies, and facing hostile locals. Moreover, their interpreters were increasingly unreliable, and they were in a volcanically active region.

No previous expedition had provided such precise data about lands so far from the Mediterranean. Its motives are unclear: an initial extravagant voyage of settlement seems to have evolved into one of exploration. Hanno may have been commissioned to encircle the continent, but was unable to do so.[6] A search for metals may have also been part of the project.[7] Greeks probably knew about the expedition within a generation or two, and the account of Hanno remains the most detailed report about the west African coast before the Portuguese explorations of the fifteenth century. There were other Carthaginian expeditions, but they are much more poorly known, and none has the benefit of an early Greek translation of its report.

About the same time that Hanno went south, Himilko went north, although his voyage is not documented before Roman times.[8] Himilko went beyond the extremity of Europe, and three references to him in the *Ora maritima* of Avienus, gathered from ancient Carthaginian sources, are all that is known about his cruise.[9] Substantive details are thus few, but imply activity in coastal northwest Europe and that he may have been the first from the Mediterranean to visit Ireland.[10] Himilko spent four months on his expedition (in contrast to Hanno's single month), which allows for a wide-ranging journey, as far as Ireland and perhaps even to the Azores, but this all remains speculative.[11]

A third Carthaginian, Mago, may have attempted to establish the routes across the Sahara. He claimed to have crossed the desert three times without water, which, if not outright exaggeration, demonstrates an astute knowledge of the journey from oasis to oasis.[12] Whether this is the same Mago who appeared at the court of Gelon of Syracuse (reigned *c.*491–478 BC) and claimed that he had circumnavigated Africa cannot be proven:[13] here "Mago" (or "Magos") may not be a proper name but a Persian magus. This is remindful of the contemporary Persian expedition led by Sataspes, who was ordered by Xerxes to make the circumnavigation of the continent.[14] The journey was not completed, with the result that he was executed by the king, but it is rare evidence of Persian activity in the western Mediterranean and the Atlantic. Two points about Sataspes' cruise are of interest: like Kambyses a generation earlier, he encountered "small men," possibly pygmies, and the reason that he failed was because his ship became stuck, perhaps caught in oceanic vegetation or a strong ebb tide—phenomena little known at this time—or even a river outflow. He may have been among the first to document the oceanic tides.

For a long time, Greek knowledge of these expeditions was minimal. Those of Himilko and Mago seem hardly to have been remembered, and even the Greek translation of Hanno's report was probably only a small part of the original. The Carthaginians discouraged any expeditions other than their own into these regions: in fact it was said that they would drown anyone who approached the Pillars, and as early as the beginning of the Roman Republic a treaty between Rome and Carthage excluded the shipping of the former from areas claimed by the latter.[15] A Greek who seems to have made it to Kerne in the mid-fourth century BC was told that no one could sail any farther south due to local conditions, perhaps a deliberate Carthaginian attempt to dissuade any Greek investigations in the region.[16] The unknown author was able to report on trading conditions at Kerne, yet, other than the Greek summary of Hanno's report, there was little Greek awareness of Carthaginian activity in the Atlantic until Carthage was conquered by the Romans in 146 BC. At that time its libraries were saved and works were translated into Greek or Latin, thus revealing the true extent of Carthaginian exploration. The historian and explorer Polybios, who was present at the fall of Carthage, seems to have been the one to implement this process.[17]

Geography in Greek Literature of the Early Fifth Century BC

As geographical knowledge expanded, it began to work its way into the writings of authors whose main interest was neither geographical nor historical. Pindar, who was active in the first half of the fifth century BC, is remembered today for his collection of songs commemorating the victors at the four great athletic festivals of Olympia, Delphi, Nemea, and Isthmia. He was Boiotian in origin but received commissions from throughout the Greek world, from Thrace to Kyrene and Sicily.[18] How much he visited the places that he wrote about remains uncertain—it is always a problem to determine whether an author used autopsy or not[19]—but his writings, as well as those of his contemporary Aeschylus, demonstrate a more extensive geographical knowledge than any previous Greek authors who were not writers specifically on the topic. His utility to geography is shown by the fact that Strabo cited him nearly 30 times, often from poems that have not survived.

Pindar's geographical notices are expressed in terms of mythology. He mentioned the Pillars of Herakles several times as a metaphor for extreme achievement: Theron of Akragas, who won the chariot race at Olympia in 476 BC, "grasped the Pillars of Herakles." Aristokleides of Aigina, winner of the pankration at Nemea, went "beyond the Pillars of Herakles, farther into the inaccessible sea."[20] Several times Pindar referred to the Hyperboreans, those "Beyond the North," a term that first appears in the Homeric *Hymn to Dionysos*, as one of the remote peoples that the god reached,[21] and who had been the goal of Aristeas in his journey north of the Black Sea.[22] The word "Hyperborean" is not an indigenous name but a Greek descriptive adjective turned into an ethnym (never a toponym), perhaps representing a faint undefined understanding that there were far northern or eastern peoples, something that Homer may have loosely comprehended.[23] To Pindar, the Hyperboreans were another metaphor for great athletic achievement as well as remote peoples among whom the fame of the successful athlete would be heard, as far distant as the source of the Nile. It was only with Herodotos that there was any scholarly attempt to analyze them.[24] Yet, in his citations of the Hyperboreans, Pindar demonstrated that the terminology and concepts of geography were entering common literary diction.

Pindar's brief account of the return of the *Argo* connects the voyage with the foundation of Kyrene, in a poem for the local citizen Arkesilas, who won the chariot race at Delphi in 472 BC.[25] Pindar had the Argonauts come to the Oracle of Ammon in the western desert of Egypt, as well as to Thera, the mother city of Kyrene, the earliest known example of manipulating geographical data for political purposes, something that the Argonaut tale was particularly susceptible to. The account is probably based on sources at Delphi or Kyrene.

The tragedian Aeschylus, whose seven extant tragedies are the earliest survivals of the genre, was more astute geographically than Pindar. Klytaimnestra's speech early in the *Agamemnon*, recounting a series of signal fires from Troy to Argos, is solidly geographical, and probably a record of early long-distance communication.[26] The route is perfectly reasonable, extending from the heights of Mt Ida near Troy to the peaks of Lemnos, Athos, the watchtower of Makistos in Thessaly, across the Euripos to Mt Messapion in northeastern Boiotia, Kithairon, Aigiplanktos in the Megarid, and finally arriving at Arachnaios, the mountain east of the Argolid. Unlike Pindar's use of metaphorical remote toponyms and ethnyms, this is a plausible list of the highest summits between Troy and Argos. Moreover, the itinerary in the *Persians* that describes the return of the Persian fleet is similar, perhaps based on what was common knowledge in post-war Athens, as is the account of the conquests of Dareios.[27]

In the *Prometheus Bound*, a speech by Prometheus presents a remarkable outline of what was known geographically about the northern and eastern portions of the inhabited world.[28] Prometheus describes to Io how she will wander from the Caucasus to Egypt. It is jumbled geographically, but over a dozen toponyms and ethnyms are preserved. There are certain correspondences with the *Circuit of the Earth* of Hekataios of Miletos, such as the Chalybians, who lived in Anatolia, but with details not found in earlier sources, including their expertise at iron working.[29] Some of the toponyms are otherwise unknown, such as the Hybristes River—if a toponym and not merely a descriptive term—but they may have come from lost parts of Hekataios' work. Aeschylus knew of the two continents of Asia and Europe, but had the common contemporary uncertainty regarding their division. As one would expect in a drama, the narrative is a mixture of contemporary geographical knowledge and the mythic heritage, but is demonstrative of how the new geographical data had begun to penetrate beyond the specialist literature.

Aeschylus' *Prometheus Unbound* is known from only a few fragments, but these reveal that it also had geographical details. Four are from the two geographical authors Strabo and Arrian.[30] In this play Aeschylus divided Europe and Asia at the Phasis River, the point of view that was falling out of favor in his day. The most interesting fragment concerns the Stony Plain of Liguria (modern Plaine de la Crau in the Rhone delta), where rocks seemed to move spontaneously, one of several places in the world where this happens, something that is generally attributed to wind and water action.[31] Aeschylus' interest was mythic, not scientific—Herakles was said to have been there—yet the citation is a rare western locale for the tragedian, and shows that his geographical knowledge extended from the Caucasus to the Keltic world. Most importantly, Aeschylus was the first popularizer of geographical data. Previous descriptions, such as those by Skylax or Hekataios, were scholarly treatises, but the plays of Aeschylus were intended for the general public, and the demographic cross-section that attended the Athenian dramatic festivals could now learn about the far Caucasus or the route to Troy.

Someone who did have a scientific interest in the earth was Xanthos of Lydia, who lived during the reign of Artaxerxes I of Persia (465–424 BC). In a passage preserved by Strabo, which was transmitted from Xanthos through Eratosthenes, he reported on the effects of a drought that he had seen (location not specified), as well as seashells far from the sea, which he observed in Armenia, Matiene (in the modern Turkish–Iranian border zone), and Lower Phrygia.[32] This is a broad sweep of territory—all the way from central Anatolia to what is now northwestern Iran—that he called a "plain," an odd term for these rugged uplands, but which may indicate that he had made several different observations at separated locations. There was also a salt lagoon somewhere—perhaps more than one—all of which suggested to Xanthos that the region had once been a sea. He was not the first to examine the question of how land was formed—Xenophanes of Kolophon had done the same a century earlier[33]—but Xanthos seems to have considered the issue more globally. He also believed that the contemporary drought provided a parallel for the drying up of the primeval sea he had postulated. An interest in drought may have come from Empedokles of Akragas, whose biography he wrote and who may have been one of his teachers.[34] Xanthos, like Empedokles, was also interested in tectonic matters. In the Katakekaumene ("Burned Up Territory"), east of Lydia, he observed the strange changes that took place in that region of contemporary vulcanism.[35] Unlike the matter of the seashells, no scientific explanation is preserved. Nevertheless both Eratosthenes and Strabo valued Xanthos' scientific ability at investigating issues regarding the formation of the earth, a tangential part of geography.

Herodotos

Much of the topographical and geographical data of the previous centuries culminated in the *Historiai* (*Researches*, or *Histories*) of Herodotos of Halikarnassos, composed in the second half of the fifth century BC, the first extended prose work in Greek. Herodotos was not a geographer but nevertheless he had a deep interest in many topics that border on geography, including geology, topography, and ethnography. He was born in the early fifth century BC, at the time of the Persian Wars. He lived in Halikarnassos, Samos, and Athens, and was involved in the foundation of the Athenian settlement at Thourioi in southern Italy (444 BC), where he spent the rest of his life. The latest comments in the *Histories* are from the beginning of the Peloponnesian War in the 420s BC, and presumably he died shortly thereafter.[36] He was thus well placed to learn about eastern Greece, Athens, and southern Italy, and his work reveals this, but he traveled far beyond these regions. At a minimum, he visited the north shore of the Black Sea and the Skythian hinterland, the Levant and probably as far inland as Mesopotamia, and had an extended stay in Egypt, going as far as the First Cataract of the Nile.[37] The reason for such extensive travels—estimated as covering nearly three million square miles[38]—remains obscure, but by his own account some of them were for reasons of pure research, and he was perhaps the first Greek to make such journeys.[39]

Herodotos' main theme was the conflict between the Greeks and barbarians, a term loosely applied but centered on the Persians, with a basic chronological extent from the rise of Lydia in the early seventh century BC to the final Greek defeat of Persia in 479 BC. His purpose is stated at the beginning of his work:

> The researches of Herodotos of Halikarnassos are set forth here, so that what happened will not fade away from men because of time, and so that the great and marvellous deeds—both those of the Hellenes and of the barbarians—will not lack renown, especially the reason that they made war against each other.[40]

These opening words—which recall the beginning of Hekataios' *Genealogiai*—allow a wide range of topics, a device that Herodotos used as thoroughly as possible. He was also aware of Hekataios' *Circuit of the Earth* and seems to have had it in mind more than once, and cited Homer, Skylax, Pindar, and Aeschylus.[41] Most of his sources, however, were oral or autoptic, sometimes based on public inscriptions.[42] The importance of Herodotos to geographical knowledge cannot be overestimated, as he brought together all that had been learned about the inhabited world up to his time. He remains the first extant Greek author to write about the entire extent of the known world, from the Keltic lands in the far west near the sources of the Istros[43] to India in the east, and from the Skythians in the north to the Aithiopians and sub-Saharan Africa in the south.[44] Without realizing it, Herodotos outlined much of the totality of the inhabited world as it was to be known in classical antiquity.

He is remembered today for his lengthy ethnographic digressions, which are without precedent in extant Greek literature. He may have relied on Hekataios for this concept, as well as local city histories within the Greek world, such as those on the foundations of Kolophon and Elea by Xenophanes,[45] or the *Foundation of Cities* by Charon of Lampsakos and the *Periegesis of the Inhabited World* by Dionysios of Miletos.[46] These works have not survived, and with the exception of those of Xenophanes, are close to Herodotos chronologically, so the question of influence is difficult to determine, but they may have helped establish the genre that he used and developed. Ephoros of Kyme, a century after Herodotos, believed that he had been influenced by Xanthos of Lydia,[47] and while it is certain that both Herodotos and

Xanthos discussed similar topics, especially the formation of the earth,[48] actual chronological precedence cannot be shown.

The major ethnographic digressions in the *Histories* include Babylonia, the Massagetians beyond the Caspian Sea, Egypt, the Aithiopians, and the Skythians.[49] In some cases, Herodotos' material was not superseded until modern times. There are many problems with his accounts, despite their great value, most notably that there is no evidence that Herodotos spoke or read the indigenous languages: thus he had to rely on local informants and their biases, who, like tour guides everywhere, exaggerated the virtues of their homeland and translated public inscriptions to their own advantage.[50] Herodotos felt that he was under no obligation to believe everything he had heard, but it was necessary to write it all down.[51] His ability, therefore, to provide a large amount of data—even if he believed it to be incorrect—is particularly important in regard to geography. For example, his error about the Alps and Pyrenees shows the actual state of knowledge in the mid-fifth century BC, when these places were only reported through hearsay.[52] His rejection of the Phoenician circumnavigation of Africa is based on the detail that provides the very proof of the voyage: the position of the sun.[53] His several theories about the cause of the flooding of the Nile include the correct one—snow melt—which he rejected (although finding it plausible) because it was thought that there could not be mountains in central Africa, since the equatorial regions were believed to be hot and dry.[54] Yet others of the same era, such as Euripides, Anaxagoras, and Demokritos, had some idea that there might be mountains in the south, but there are few specifics about their ideas on the matter.[55] Nevertheless, these are the first hints of the mountains that do exist in the regions around the source of the Nile, marking the beginning of a lengthy controversy about the nature of the land around the equator.

Herodotos, along with Xanthos of Lydia, was among the first to consider geological issues. In Egypt he noted that the delta was topographicaly anomalous (extending far out to sea) and that there were seashells in the uplands.[56] Moreover, the land of Egypt was unlike its surroundings, created by mud pouring forth from Aithiopia, and different from the sandy red soil of Libya to the west or the clayey stony soil of Syria. He also knew that the mud was pushed out to sea, as far as "a day's run from land," which by his own definition would be 60,000–70,000 *orgyiai,* or about 70 miles.[57] This mud could be sounded at a depth of 11 *orgyiai,* or about 70 feet. Yet these were merely observations that intrigued Herodotos, and no theoretical explanation for them was advanced.

Herodotos' loquaciousness made him a valuable source for the state of geographical knowledge by the latter fifth century BC, at the time of the beginning of the Peloponnesian War. His skill at observation and deep interest in the ethnography of remote areas provided much data for later, more scientific, geographical analysis.

Geography in Historical Writers of the Latter Fifth Century BC

By the end of the fifth century BC, geographical digressions had come to be a regular part of historical writing, following the model of Herodotos. Thucydides, for example, as preface to his account of the misbegotten Athenian invasion of 415 BC, included a report on Sicily, which is primarily historical but contains a certain amount of topography. Another topographical excursis is about the Athenian Akropolis, and there is an ethnography of the Thracians, each inserted into its relevant place in the history of the Peloponnesian War.[58] On occasion Thucydides seems to have relied on a *periplous*, most notably in his account of the location of Epidamnos in northwest Greece ("Epidamnos is a city on the right as one sails into the Ionian Gulf").[59] Thucydides' accounts do not have the exoticism that one finds in reading Herodotos, but demonstrate that history and topography had become linked.[60]

An older contemporary of Thucydides, Antiochos of Syracuse, may also have been in this tradition. Only a handful of fragments of his works survives, which were titled *Sikelia* and *On Italy* (a toponym at this time referring only to the southern part of the peninsula), and which demonstrated an understanding of the topography of those regions.[61] Over half the fragments were preserved by Strabo.[62] Antiochos may have been the first Greek to discuss the founding of Rome, a tale that in its familiar version has south Italian connections.[63] He certainly expanded Greek knowledge of Italy from the coastal regions known to Herodotos into the interior and to the north, beyond the Bay of Naples, although, as with any fragmentary author, the true extent and tone of his works cannot be determined.

Ktesias of Knidos is problematic as an historian, but nevertheless is important in the development of geographical data. He was physician to the Persian king, Artaxerxes II, at the end of the fifth century BC, and later served as a Persian diplomat to the Greek world.[64] He had every opportunity to learn about Persia and the east, but how much he made use of this has long been debated, and he seems to have had a taste for exaggeration, the spectacular, and the fantastic, and to have been driven in part by a need to contradict Herodotos.[65] Yet this hardly means there is nothing of value in his writings.

The surviving fragments of Ktesias' *Indika* are largely preserved in a summary by Photios, patriarch of Constantinople in the ninth century AD. It seems to have been the first treatise devoted solely to that part of the world, preceded only by Herodotos' chapters on the region.[66] Ktesias' account is full of marvels—as were all other Greek works about India—yet generally the material is different from that of Herodotos. The sources remain uncertain, but for the most part they were obtained at the Persian court. One reads of a fountain of liquid gold, large serpents, the man-eating *martichora* (probably the tiger), the Dog-Headed People, a worm that can kill an ox, and many other strange phenomena, some of which may not seem as unusual today as they were in Ktesias' time. The extant account is heavily slanted toward flora and fauna, with some ethnography, but no history and practically no topography. The few toponyms are difficult to identify.[67] The *Indika* should not be dismissed, and while it contains questionable elements, it is important in being the most complete Greek account of

India before the time of Alexander the Great, thus extending the horizon of the inhabited world. Despite his sense of the exotic, Ktesias may have been a significant player in the development of geographical writing. His other geographical work, *Periodos* (*Going Around*), survives in only six fragments, which reveal that it included Egypt, the Black Sea, Anatolia, central Italy, and Libya, but nothing is known about its format other than it was at least three books long. Again there is an emphasis on the fantastic, since he mentioned the Skiapodes ("Shade-Footed People"), Libyans whose feet were so large that they could use them as sunshades.[68]

In 401 BC, Cyrus (generally known as Cyrus the Younger), the satrap of western Anatolia, decided to overthrow his brother, Artaxerxes II, the Persian king, who was also Ktesias' employer. With the assistance of a large number of Greek troops, he mounted an invasion of Persia. The attempt failed, largely because Cyrus was killed that autumn. Yet the expedition remains one of the best-known events of ancient history because of the exhaustive treatment by the Athenian Xenophon, who was a participant. His *Anabasis* not only recounted the journey with Cyrus but also the difficult return of the Greeks after his death.

The *Anabasis* ("Going Up," a relatively new term in Xenophon's day and referring to the journey up from the coast into the interior of Asia) describes the entire expedition from its mobilization at Sardis in the spring of 401 BC to the return of the Greek forces to western Anatolia the following year, and the eventual disposition of the army in 399 BC. The forces headed inland from Sardis, and from the beginning Xenophon (whose initial role seems to have been minor) kept a log of the journey, or at least had access to the official Persian one. Rather than follow the Royal Road to the east, which Herodotos had outlined, where movements might be more easily detected, Cyrus kept to the south in the more rugged territory of Pisidia and Kilikia, reaching Dana (Tyana).[69] From here he turned south and passed through the Kilikian Gates down to the coast and the ancient and famous city of Tarsos.[70] He then headed around the northeastern end of the

Mediterranean to the trading center of Myriandos,[71] from where he went directly east to the Euphrates at the river crossing of Thapsakos, which is not located with certainty but would become a major point in Eratosthenes' grid of the inhabited world.[72] From there the expedition headed down the Euphrates, eventually engaging the forces of Artaxerxes II at Kounaxa, not far from Babylon. Here Cyrus was killed and Artaxerxes wounded; the king was attended by his physician, the historian Ktesias.[73]

The death of Cyrus placed the Greek army in a quandary. They were far from home in hostile territory, and the purpose of the expedition had evaporated. The Persians were initially friendly, but soon turned hostile and executed the Greek commanders. The Greek troops were on their own. Eventually Xenophon emerged as the leader, and became responsible for getting them home. The return was much more difficult than the advance, as they retreated north into rugged territory populated by unfriendly mountain people. Although Xenophon recorded the march in detail, it is impossible to determine its exact route until, after many encounters with the locals, they eventually sighted the Black Sea (with their famous cry of "the sea, the sea")[74] and a few days later reached the ancient Milesian settlement of Trapezous (modern Trabzon). From here it was merely along the coast back to the Greek heartland at the Bosporos.

The *Anabasis* is an unusual piece of literature, a mixture of history, autobiography, and a travel account. It was written in the third person, and the reader would never know that the Xenophon who played such a prominent role in the retreat was the author. From the point of view of geography and topography it generally does not go beyond regions known to Greeks of the period, except in some of the more remote mountain areas of eastern Anatolia, and the military quality of the treatise is always paramount. Nevertheless Xenophon recorded geographical and ethnographic details previously unknown, such as the Kardouchians and the Taochians (not even mentioned by Hekataios or Herodotos), the pernicious effects of the local honey near Trapezous, and the customs of the Mossynoikians who lived along the Black Sea.[75] And Xenophon was the first Greek to travel over the extremely rugged country from Mesopotamia to the Black Sea and to seek the source of the Euphrates, something Herodotos knew nothing about except that it originated in Armenia.[76] But the expedition did not realize that this took them into the high country in midwinter, with deep snow and a violent north wind.[77] In some ways the expedition of Xenophon is a precursor to that of Alexander the Great 70 years later, and even those of the Romans, demonstrating the expansion of the geographical horizon through military campaigns. But the experience in the snows of Armenia may have been one of the reasons Strabo wrote that field commanders ought to know geography.[78]

In addition, geographical knowledge began to penetrate other scholarly disciplines. A good example is the treatise *Airs, Waters, and Places*, in the Hippokratic corpus.[79] It is a brief essay on the effects of climate on health, contrasting the people of the far north (the Skythians) with those of the south (the Egyptians), thus comparing the northern and southern extremities of the inhabited world. The work is not up to date on geographical theory, as it presumes only two continents (placing Egypt in Europe), but has a wide geographical range, with a detailed discussion of the region of the Phasis River and that north of the Black Sea. Its tone subordinates anthropology and geography to medicine, but it demonstrates the usefulness of geographical knowledge to medical theory.

The Zones and the Size of the Earth (Map 5)

Theorization about the nature of the inhabited world became more intense during the sixth and fifth centuries BC, given the large number of geographical facts flowing from the extremities into the Mediterranean heartland. The Pythagoreans had suggested that the world was a sphere, and even though this was not fully accepted as late as Hellenistic times, it provided a starting point. There was evidence that different places experienced different celestial phenomena, most

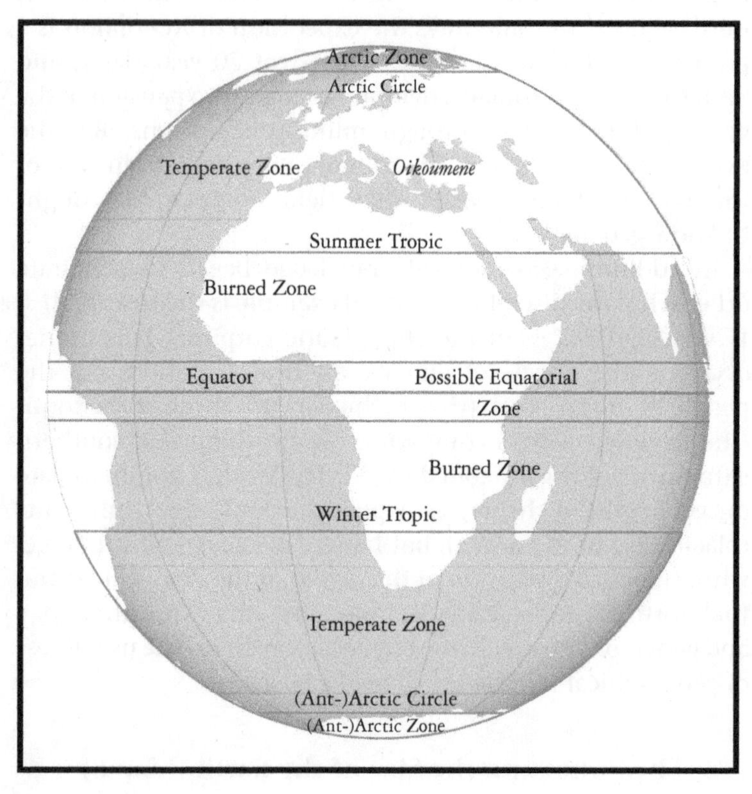

Map 5. The zones.

notably the varying length of days as one went north or south, and also, roughly speaking, that it was colder to the north than to the south. These observations were used as the practical basis for theories about the overall surface of the earth.

Parmenides of Elea (in southern Italy) was allegedly a student of Xenophanes of Kolophon and had Pythagorean connections.[80] He was active in the early fifth century BC, and was perhaps the first to consider scientifically the nature of the whole earth. Building on the knowledge that the climate differed with latitude—although such concepts were not as yet formalized—he divided the earth into five zones.[81] The zones are a concept that is more complex than it seems at first glance, for the idea requires a blending of natural philosophy, astronomy, and topography, a mixture of disciplines that became essential to geographical thought.[82] Zonal theory is based on a spherical earth, which the natural philosophers had provided, and also a projection of the circular path of the heavenly bodies onto the surface of the earth, particularly that of the sun, as well an understanding of the types of shadows cast at different latitudes. In addition, the theory assumes an arctic circle—the circle of stars that never sets—so called because it was determined by the constellation Arktos, the Bear, which had been known since earliest times and had been of major importance to sailors.[83] This celestial arctic circle was also projected onto the surface of the earth.

Parmenides' five zones were two arctic ones, two temperate ones, and a single "burned" zone, which was essentially twice the width of the others. This zone was between the tropics—the "turns"—the imaginary lines on the earth below the path of the sun at the solstices, above which the sun was seen to turn and reverse its north–south movement, and which were defined by the longest and shortest days of the year. The extant source, Strabo (from Poseidonios) has combined the views of Parmenides, Aristotle, and Poseidonios, as well as his own, but if Parmenides used the word "tropic" (*tropikon*), it was new terminology.[84] By Strabo's day the double width of the burned zone had become controversial, and late Hellenistic arguments center on this issue, but Parmenides' temperate and arctic zones have remained standard. The nomenclature shows the mixture of influences: one zone has an astronomical name and the others are based on climate. This theory of zones is the first to conceptualize the surface of the entire earth, and is notable because neither Parmenides or, one presumes, anyone else from the Mediterranean, had ever been in the arctic zone, which begins hundreds of miles north of the Black Sea beyond the sources of the rivers flowing into it. Yet it was nevertheless possible to assume its existence. On the other hand, the tropic—and thus the burned zone beyond it—was more accessible, since it lay at Elephantine on the upper Nile, which Greeks had known from at least the fifth century BC.[85]

Implicit in the theory, but not as yet verbalized, was the location of the known portions of the earth within its whole. Moreover, the idea of zones presumes a southern half of the earth, beyond human habitation, with a second temperate zone and then a second and southern arctic zone that were opposite to the northern ones.[86] There was probably no theorization as yet about the size of the earth, but the sense of zonal divisions and the use of astronomy to locate positions on its surface would eventually make this possible. In fact, Demokritos of Abdera, later in the fifth century BC, saw the inhabited world as an oblong, with its length (east–west) one-and-one-half times its width (north–south), which would seem to place it totally within the temperate zone.[87] The belief that the inhabited world was longer than wider is the probable origin of its visual orientation with north at the top and east and west to the right and left, which became standard from antiquity until today (with some exceptions in medieval times).[88] These theories were further refined by Plato, who spent time in Italy and Sicily and thus was exposed to Pythagorean and other western thought. He believed that the earth was exceedingly large (using a new word, *pammegas*), and that the inhabited portion—lying between the Pillars of Herakles and India—was only part of it. Moreover, there were many other people in places beyond these known limits.[89] He also began to consider the significance of "above" and "below" on the sphere of the earth. If one were to travel from one pole to the other, it would be seen that those in the far south would have their "feet opposite," or *antipous*, another new word, more familiar in the plural as Antipodes, which eventually became a hypothetical toponym.[90] There is no evidence that Plato provided any specifics for these ideas, but it was beginning to be realized that the areas of human habitation were only a small portion of a large earth. If a comment by Horace is to be taken literally, Plato's associate, Archytas of Taras, a Pythagorean, may have been the first to attempt a determination of the size of the earth.[91]

Eudoxos of Knidos, a student of Archytas, was active in the first half of the fourth century BC.[92] He was primarily a mathematician and astronomer—one of the first to consider planetary motions—yet he wrote a geographical work, *Circuit of the Earth*, its title reflecting that of Hekataios of Miletos a century and a half earlier.[93] Strabo believed that Eudoxos was a notable figure in the history of geography, and there is no doubt that his contributions were significant, although the details of his treatise remain elusive.[94] Nevertheless it was at least seven books long, perhaps the most substantial work on geography to date. Much of the extant material is about Egypt, where Eudoxos lived for some time, but the work extended to the Skythians and beyond. It may have been limited to the eastern Mediterranean and points east, although this reflects the extant fragments.[95]

Eudoxos also considered the dimensions of the known inhabited world, which by now extended from the Pillars of Herakles to India and from north of the Black Sea to Aithiopia, refining Demokritos' figures to propose that this meant its length (east–west) was twice its width (north–south).[96] Eudoxos may also have been responsible for the profound idea—which would influence exploration for the next 2,000 years—that one could reach India by sailing west from the Pillars of Herakles. The earliest extant source for this belief is Aristotle,[97] a younger contemporary of Eudoxos, but the concept fits so neatly into Eudoxos' theories that he was probably the originator, and Aristotle may have heard it directly from him.

Such thoughts would lead naturally to consideration of the size of the earth. The earliest known figure for its circumference is 400,000 stadia, again from Aristotle, and attributed merely to "the mathematicians" but perhaps also part of Eudoxos' treatise.[98] Eudoxos also had further thoughts about who might live in the southern temperate zone, for whom the seasons would be reversed.[99] He refined Parmenides' vague thoughts about the terrestrial zones, developing the concept of the *klima*.[100] The word means "slope," reflecting the perceived slope of the spherical earth from the equator to the poles. By Hellenistic times it had become the standard word for "latitude," but it is not certain that Eudoxos had this degree of precision.[101] Nevertheless with the zones and the realization that people lived in only a small portion of an immense earth, it became natural to see it in terms of east–west *klimata*: the emergence of a sense of latitudes.

Aristotle and Geography

Astonishingly, Aristotle does not seem to have written a geographical treatise, perhaps an indication that geography was still on the fringes of mainstream scholarship, yet geographical data are scattered throughout his other works, especially the *Meteorologika*, whose primary topic is weather phenomena.[102] The section that is most geographical is a lengthy discussion about rivers, primarily about their water supply, which treats issues of flow, rain, evaporation, and the role of mountains in creating the moisture that ends up in rivers.[103] Included are various comments about the mountains and rivers of the world. The list is confused—the Paropamisos (Hindu Kush) is called Parnassos (a mountain in central Greece)—yet it reveals a breadth of knowledge about the rivers east of the Caspian as well as those in western Europe and in the north of the continent. There is also the mention of the mountains at the source of the Nile. The Chremetes, an African river flowing into the Atlantic, is otherwise only known from Hanno's *periplous*, a suggestion that Aristotle may have been familiar with that treatise.

Otherwise, he provided extensive information about the thoughts of his predecessors, from Thales to Eudoxos, including many of the important bits previously noted, such as the first extant figure for the circumference of the earth. He elaborated Eudoxos' thoughts about sailing from the Pillars of Herakles to India, suggesting (rather strangely) that the presence of elephants in both northwest Africa and India (in other words, at the extremities of the inhabited world) was proof of a connection between these regions.[104] On a practical level, he seems to have accepted the idea that the inhabited world was encircled by a continuous External Ocean, and he had the sense of a distant sea into which the large rivers of northern Europe drained, perhaps the first glimmerings of the Rhine, Elbe, and Vistula. That he had some knowledge of Europe beyond the Alps—although his source is unknown—is demonstrated by one new toponym—the Arkynian Mountains—which were north of the Istros (Danube), and are presumably the forested uplands beyond that river.[105] Aristotle also stressed Eudoxos' view that the length of the inhabited world was more than its width, providing a ratio of 5:3 for length (Pillars of Herakles to India) to width (Aithiopia to Lake Maiotis), the first attempt to construct relative distances across the inhabited world. He based his conclusions on travelers' reports, which would remain the standard means of obtaining geographical data throughout antiquity. He also defined the boundaries of the inhabited world: in the east and west there was the External Ocean, in the north regions too cold for people to live in, and in the south excessive heat. Finally, he established the term for the inhabited world, *oikoumene*, a new word: whether or not Aristotle invented it is not clear.[106] His contemporary, Demosthenes, used it to characterize the civilized, or Greek, world in contrast to those not civilized (to Demosthenes, the Macedonians), a definition that, needless to say, would soon become obsolete.[107] Yet as a geographical term for the inhabited portion of the earth, as opposed to that which was uninhabited, it would remain standard diction throughout antiquity.

New Geographical Knowledge of the Early Fourth Century BC

The first half of the fourth century BC was a fertile period for geographical theorization, yet there was relatively little exploration or toponymic expansion in the period between Xenophon's expedition and that of Alexander the Great, 70 years later. There is some evidence for additional Greek knowledge of the Atlantic coast south of the Pillars of Herakles, which had been previously explored by the Carthaginians. Although the early Greek investigations of Euthymenes were kept secret by the Massalians and were not generally disseminated, a few Greeks may have penetrated into the area: a cluster of Central Greek names (Kephisos, Kotes, and Pontion) is a memory of someone from that region who reached the Atlantic coast of Africa. The location of these places is uncertain but they are probably not far south of Cape Spartel, the northwest corner of the continent, demonstrating that this unknown Central Greek did not go a great distance along this alien and hostile coast.[108] The source for the names is the *Periplous* of Pseudo-Skylax, a description of the coast of the Mediterranean and the Black Sea, with inclusion of a small part of the adjacent Atlantic littoral. In late antiquity, this was believed to be the work of Skylax of Karyanda, the explorer of Dareios I of Persia, but it actually dates from the early 330s BC, reflecting the world just before the accession of Alexander the Great in 336 BC.[109] It seems to be a compilation of earlier data and not an eye-witness report, by an author whose identity is unknown. Although much of it describes coasts well known to contemporary Greeks, at the end of the work is a single chapter on the Atlantic coast of Africa, and which provides new information, the first extant Greek account of this region. As expected, seamanship and mercantile information dominate the report: there are great reefs beyond Cape Hermaia, and a rich trading process took place at the Carthaginian outpost of Kerne, where Attic pottery and other items were available, something supported by archaeology.[110] The area was still very much under Carthaginian control in the mid-fourth century BC, but nevertheless a few Greeks and some Greek trade objects had penetrated the region.

Also in the fourth century BC, Greeks became aware of Carthaginian reports of a fertile island in the Atlantic, discovered at an uncertain date, when (inevitably) ships were off course due to a storm.[111] This seems to be one of the Madeira islands, which lie 360 miles from Africa and 425 miles from the Iberian mainland. The island remained outside the mainstream of Greek knowledge, and in fact came to be identified with the mythical Blessed Island of Homer. It was one of the more remote places known to classical antiquity.[112] The Azores, 480 miles farther northwest, may also have been visited by the Carthaginians, and Greek coins from Kyrene have been found on the island of Corvo, but the uncertainty of the published report and inconclusive archaeological investigations mean that it is impossible to conclude whether the islands were visited—in anything beyond a brief, accidental manner—in ancient times.[113]

There seems to have been some additional awareness of Europe north of the Alps, although the data are especially vague. Herodotos believed that the source of the Istros (Danube) was at the town of Pyrene, a misapprehension of the topography of central Europe based on reports from near the mouth of the river.[114] Aristotle, however, knew of the forested lands to the north of the Istros and the great rivers flowing toward the North Sea and Baltic, without reporting their names. The Rhine (Rhenos in Greek, Rhenus in Latin) is not cited with certainty in extant literature until the time of Julius Caesar, but his Hellenized spelling suggests a Greek source, probably Poseidonios.[115] The Vistula may have been known as early as the latter fourth century BC, if Pytheas of Massalia got that far.[116] It lay on the ancient amber route, and would have been approached from the Mediterranean before the Greeks learned about the more western rivers of the north. Other northern rivers are identified only later.

A certain Timagetos wrote *On Harbors*, probably in the early fourth century BC.[117] Both work and author are obscure, but he placed the source of the Istros in a Keltic lake, perhaps early knowledge of the Swiss lakes. The extent of the work is unknown, yet five of the six existing fragments concern the course of the Istros (in the context of the return of the Argonauts), and Timagetos may have had access to material about northern Europe that was also used by Aristotle.

There was further interest in the source of the Nile, a continual intellectual problem that lasted until the nineteenth century. The mountains of central Africa had been uncertainly known since the fifth century BC, and Aristotle was the first to record a name for them, Argyos, or Silver. This was eventually eclipsed by that provided by the geographer Ptolemy: Selene,[118] more familiar in its anglicized version, the Mountains of the Moon, the great goal of nineteenth-century explorers searching for the source of the Nile. There is also the first hint of the marshes of the upper Nile, the modern Sudd of Southern Sudan.[119]

Ephoros of Kyme

By 335 BC there was an extensive amount of toponymic data available to the Greek world. The inhabited portion—the *oikoumene*—extended around the coasts of the Mediterranean and the Black Sea, and some distance into the interior. There was also comprehension, in varying degrees of detail and reliability, of what lay beyond: the Keltic lands in the west, the rivers north of the Alps and those entering the Black Sea, the Caspian and the regions beyond, India, the upper Nile, and sub-Saharan Africa. There were still many gaps and inaccuracies: the Arabian peninsula, for example, was almost totally unknown, except as the origin of frankincense.[120] The region between Persia and India remained uncertain, and there was only the faintest sense of Europe north of the Alps. Any land beyond these limits was considered uninhabitable due to either cold or heat. Nevertheless the circuit of the *oikoumene* was extensive. Parallel to this topographic knowledge was a developing theoretical structure that was based on a spherical earth, with the first rudimentary attempts to define its surface, even beyond the *oikoumene*, through analogy with the celestial sphere. There was also the realization that the earth was exceedingly large and the inhabited portion only a small part of it. Details were lacking but it was possible that there might even be other inhabited portions, somewhere in the great External Ocean that surrounded the *oikoumene*.

Yet there was no organization of these thoughts—whether theoretical or topographical—into a discipline of geography. The two paths of inquiry were being considered separately: theoretical matters were the concern of the natural philosophers and astronomers, and topography that of sailors, merchants, and governments. Much that could be called geographical had been recorded in Greek literature—far more than survives today—from Homer to the anonymous author known as Pseudo-Skylax, yet no treatise on geography had as yet been written, not even as a cohesive portion of another work. Hekataios perhaps had come the closest, but his *Circuit of the Earth*, in so far as it is known today, lacked any scientific theories or overall conception of the entire earth. Eudoxos' work of the same name may have been more scientific but topographically it was limited. The *periplooi* were two-dimensional, generally restricted to coastal toponyms and land forms. Other geographical data lie buried in works whose focus is essentially not geographical, in particular the material recorded by Herodotos.

The first significant step toward creating a discipline of geography (still without the use of the word itself) was taken by Ephoros, who came from Kyme, an Aiolian city on the coast of Anatolia.[121] Little is known about his life, but he survived into the first years of the reign of Alexander the Great (which began in 336 BC). He wrote the first universal history, in other words, one whose topic was not, like the works of Herodotos, Thucydides, and Xenophon, limited to a specific period or topic, but which included the whole of human history. In 30 books, it was probably the longest treatise as yet written, extending from the mythological period to the siege of Perinthos in 341 BC.[122] The work is not extant, but was widely cited in Hellenistic times, especially by Diodoros, Strabo, and Plutarch, and exists today in nearly 250 fragments.

Ephoros is important in the history of geography because within the context of his lengthy treatise he included a section specifically on that topic, appearing in Books 4 and 5. It is possible that these two books could have stood alone as a seminal independent work on the subject. Book 4, at least by Hellenistic times, was titled "Europe";[123] the scope of Book 5 is more difficult to determine, since the few extant fragments include the Black Sea region, Greece, and Libya.[124] Ephoros opened his geography by dividing the inhabited world into four portions, each defined by an ethnic group, and using both the prevailing winds and the positions of the sunrise and sunset at different times of the year to determine their direction from the Mediterranean world, thus for the first time joining astronomy and topography. As Strabo reported:

> in his treatise *On Europe* he says—dividing the regions of the heavens and earth into four parts—that the Indians will be toward the Apeliotes wind, the Aithiopians toward the Notos, the Kelts toward the sunset, and the Skythians toward the boreal wind. He adds that Aithiopia and Skythia are the larger, for he says that it seems the Aithiopian peoples extend from the winter sunrise as far as the sunset, and Skythia lies directly opposite them.[125]

All four of the ethnic groups cited had long been considered the most remote of peoples, and this is the earliest known attempt to conceive of the entire *oikoumene* and to create a universal set of reference points for distant places, although the scheme only works from a Mediterranean viewpoint.

It is unfortunate that fewer than 20 fragments survive from Books 4 and 5, but Ephoros' relevance to geography is demonstrated by the fact that one-fifth of the extant fragments of his entire corpus (45 in number)—many of which are quite lengthy—are recorded by the geographer Strabo. They cover a wide range thoughout the inhabited world, with extensive ethnographic discussions. There was probably more detail than ever about the north—the Keltic and Skythian world—including (without providing a name) the first specific report on the North Sea and its low-lying coasts that were often affected by storm surges and tides.[126] But the real significance of Ephoros in emergent geographical study was his ability to look at the *oikoumene* as a whole and to bring astronomical data into the ethnographic discussion. Although it would be more than a century before the discipline of geography was formalized—by Eratosthenes of Kyrene—Ephoros moved closer than anyone previous to writing a treatise on the topic.

CHAPTER 4

PYTHEAS AND ALEXANDER

In July of 336 BC, the king of Macedonia, Philip II, was assassinated at Aigai. He had spent much of his quarter-century reign putting together a state that, in one way or another, controlled most of the Greek peninsula, and was arranging an expedition against the moribund Persian Empire, in retribution for their invasion of Greece over a century earlier. Upon his death, the Macedonian army, as was its duty, elected the new king, and chose Philip's son Alexander III, who had not yet earned the title "the Great." The young king had benefited from the presence of Aristotle and perhaps his prize student Theophrastos at the Macedonian court. Alexander's accession was probably in early October, and after a campaign in Thrace as far as the Istros (which added a number of ethnyms to the toponymic map), he eventually undertook his father's planned Persian expedition, setting forth in the spring of 334 BC, just as Ephoros was completing his history.

Pytheas of Massalia (Map 6)

During roughly the same time as Alexander's campaign, which enhanced Greek geographical knowledge of the east, Pytheas, a citizen of Massalia, the most important Greek city in the west, began a journey far to the north into areas previously unseen by Greeks, increasing significantly the understanding of the North Atlantic regions as far as the edge of the Arctic.

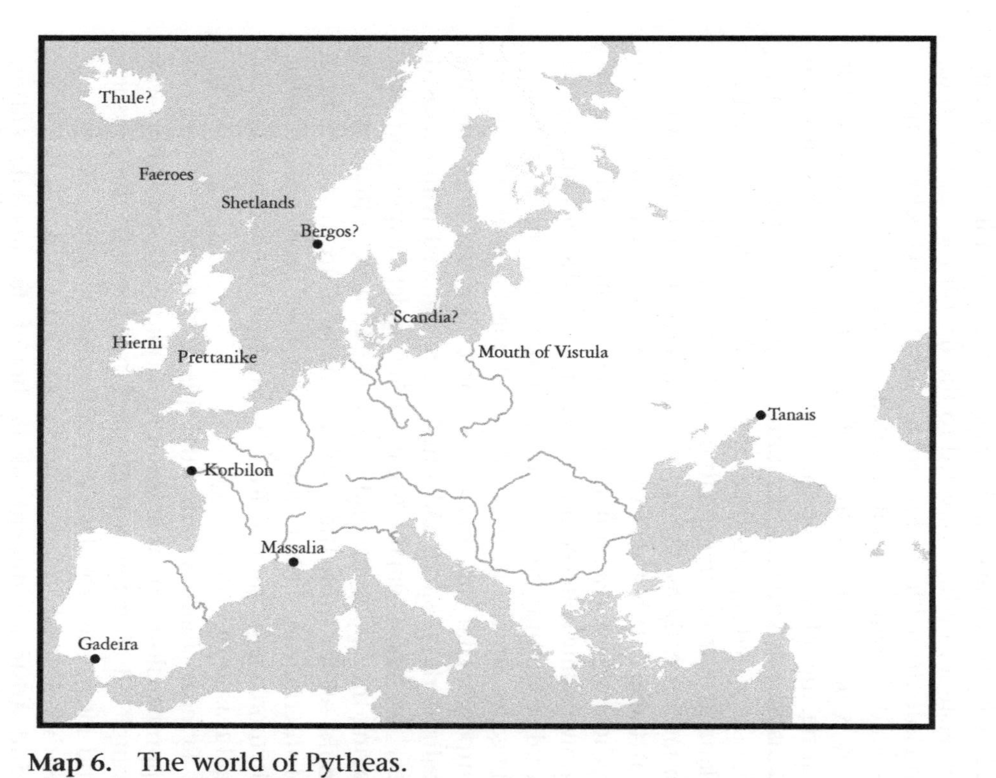

Map 6. The world of Pytheas.
Note: Some modern toponyms are used, since those of Pytheas are not always known.

Although Pytheas and Alexander were totally different personalities, together and simultaneously they made the greatest contributions to practical geography since the expansionism of the Archaic period. India, Alexander's easternmost point, had been known for two centuries, but before his time the lands between Persia and India were hardly understood; Pytheas traveled to the British Isles and beyond, and into the Baltic, regions totally new to the Greek experience beyond a few uncertainly located toponyms.

It is important for the modern reader to remember that during the 11 years of Alexander's expedition his ultimate plans were unknown and a source of worry for many, and there were repeated reports that after he returned from the east he planned to make an expedition to the western Mediterranean. In fact, as early as 332 BC, when at Tyre, he declared war on Carthage.[1] Whether the contemporary expedition of Pytheas was a reaction to a perceived threat will never be known: some believed that Alexander would attack the west by coming around the north of the *oikoumene,* and that he was allegedly already building ships on the Caspian Sea for this purpose.[2] That this was a geographical impossibility was not realized: the Caspian had long been recognized as an enclosed sea,[3] yet reports emanating from Alexander's expedition were saying the opposite,[4] and it was easy to believe that his fleet could set forth on the Caspian and end up in the western Mediterranean. Other rumors were that there was to be a circumnavigation of Africa.[5] In any case, the western Greeks had reason to worry.

Pytheas' journey falls into this environment of western apprehension about Alexander's plans. It probably was completed after Ephoros finished his history in the 330s BC, and even after the last years of Aristotle (died 323 BC), as neither showed any awareness of the Massalian. On the other hand, Aristotle's student Dikaiarchos of Messana did mention Pytheas, the earliest to do so, which suggests that Pytheas published his *On the Ocean* by the latter 320s BC or shortly thereafter.[6] It survives in about 20 fragments, most of which were preserved either by Strabo or Pliny the Elder. It is difficult to analyze the work because of the limited number of fragments and a hostile tradition about Pytheas that seems to have originated in the second century BC, perhaps with Polybios, and which was reinforced by Strabo a century later.[7] In part this was an early example of a phenomenon that lasted into the nineteenth century, in which explorers to the remote areas of the earth who reported strange phenomena were doubted or even ridiculed. Moreover, the rapid evolution of the discipline of geography in Hellenistic times made some of Pytheas' data eventually seem obsolete, and even a fantasy. He was the first from the Mediterranean to travel on the North Atlantic, and he reported many unusual things that seemed implausible to his successors. Today, with the regions that he ventured into well known, he is recognized as one of the greatest explorers of antiquity. Pytheas does not seem to have had state support, and traveled as a private individual, something that Polybios found unbelievable.[8] This may rule out any suggestion that his journey was an official response to rumors about Alexander, but nevertheless Pytheas does fit into the expanding world of that era, even if operating as an independent scholar, traveling by foot and on local fishing boats rather than as part of a government expedition.

Although controversies and uncertainties about his route will never be solved, the best suggestion is that he first went overland from Massalia, either up the Rhodanos (Rhone) to the Liger (Loire), or inland from Narbon (Narbonne) up the Atax (Aude) across to the Garounna (Garonne). The latter route reached the Atlantic at the site of modern Bordeaux.[9] He was knowledgeable about the topography of extreme northwest Europe, in modern Brittany, including the peninsula that is today the island of Ushant.[10] Up to this point he was probably in territory well known to Massalian traders.

He then crossed over to Kantion (whose name is preserved as modern Kent), which was part of the large island named Prettanike,[11] where he spent an extensive period of time, finding the indigenous inhabitants living a life of prehistoric simplicity.[12] Unfortunately the accumulation of later data on the British Isles, beginning with the report of Julius Caesar, has meant that few details of Pytheas' travels are preserved and there are no toponyms surviving from his era other than Kantion and perhaps the names of some of the Scottish islands.[13] He may have visited the Cornish tin mines (suggesting a mercantile purpose for at least the early part of the journey), and he certainly began to become aware of the long winter nights and summer days of high latitude.[14] He also made astronomical observations, and recorded tidal phenomena, something still unusual to Greeks because of their limited experience with the External Ocean.[15] In addition, he perhaps heard reports about Ireland.[16] He stayed a long time in Prettanike, possibly more than a year.

Eventually Pytheas went north from Prettanike. His route is difficult to follow, but he made a latitude calculation that suggests he went to the Faeroe Islands. Soon he reached his most famous discovery, Thule, six days from Prettanike.[17] Although the location of Thule will forever remain controversial, evidence points to Iceland, which at least in Viking times was occasionally visible from the Faeroes, a day or more to the northwest.[18] Thule, which became a mystical paradigm for the far north, was a land with a unique mixture of volcanic and glacial phenomena, situated near the frozen or solid sea. The sea in this region could also be breathing or boiling—the Greek is unclear—with the latter connected to volcanic activity. The former is metaphysical, perhaps reflected by Pytheas in his description of the "sea lung," an enigmatic concept that may reflect the rise and fall of ocean ice. At Thule the world was still being formed, and life was harsh, with the locals living even more primitively than on Prettanike.[19]

From Thule Pytheas continued to follow fisherman's routes, and ended up at Bergos, perhaps the region of Bergen in Norway.[20] There was also Scandia, the future eponym for Scandinavia, located at the southernmost point of Sweden. These demonstrate that Pytheas probably entered the Baltic, and perhaps went as far as the mouth of the Vistula.[21] The evidence is sketchy, in part because the sources are in Latin—Strabo did not describe this region—and the material has become tangled with information about the long-standing amber trade as well as

Roman operations in the North Sea. In fact, Pytheas himself may have been investigating amber sources.[22] He may even have continued overland from the Vistula to the Black Sea, following a trade route, as he was said to have covered the entire coast from the Tanais (modern Don) to Gadeira,[23] which shows that he visited the north coast of the Black Sea, but this may have been on another trip. Like many great explorers, Pytheas probably made more than a single journey, and Strabo's diction, "from Gadeira as far as the Tanais," implies heading to the east, not returning from the north.

Of additional importance is that Pytheas was not only an explorer but a scientist, and assiduously recorded and interpreted the natural phenomena that he encountered. As such he was another example, along with Ephoros, of the merging of several scholarly disciplines into geographical thought. Pytheas calculated his latitude several times, beginning with his departure point of Massalia. His methodology was to determine the height of the sun at the winter solstice: it was nine *pecheis* (about 14 feet) at Massalia, and, at three places to the north, six, four, and less than three.[24] Regardless of the validity of the data, the measurements demonstrate that Pytheas' journey was lengthy, since each had to be within the 10-day span of the solstice period. Exactly how (or when or by whom) these figures were translated into latitudes (recorded as distances north of Massalia) remains obscure, and there is the usual problem of separating Pytheas' material from that of later scholars.[25] Nevertheless this was the first attempt to determine the position of remote places on the earth and to relate them to known localities, and to have a sense of the importance of meridians. Moreover, the calculations of length of day are the best proof of Pytheas' journey to the far north, as the three measurements north of Massalia are at the latitudes of York, northern Scotland, and the Faeroes.[26]

Pytheas also theorized about the tides. He wrote that he encountered ones that were 80 cubits high (well over 100 feet), perhaps the result of a storm surge.[27] Regardless of what he experienced, he came to the important conclusion that the high tides were the result of lunar activity, and was the first to do so.[28] One of his most lasting contributions was to inaugurate the study of tidal phenomena.

Furthermore, as Pytheas went north he collected data about the peculiar meterological and astronomical conditions of the arctic. He commented on the long days and nights and even recorded a place called the Bed of the Sun, a toponym probably at Thule.[29] He observed the celestial movements at high latitudes, remarking that there was no star but merely an empty tetragon at the pole.[30]

Pytheas' achievements in astronomy, tidal theory, determination of latitudes, and his belief that Thule represented a place where the earth was not yet fully formed, are of great importance in his reputation as a geographer, and demonstrate the maturation of geography into a discipline, as well as his role in the emergent polymathic thought of the Hellenistic world. His tarnished reputation and the loss of *On the Ocean* make it easy to minimize him, but he realized that far distant places on the earth's surface could be connected to one another through latitude calculations, which eventually made possible the grid of the *oikoumene* that Eratosthenes was to develop a century later. It was no exaggeration when Fridtjof Nansen, one of Pytheas' greatest successors and someone well qualified to discuss his achievements, wrote that he

> appears to us as one of the most capable and undaunted explorers the world has seen. Besides being the first, of whom we have certain record, to sail along the coasts of northern Gaul and Germany, he was the discoverer of Great Britain, of the Scottish isles and Shetland, and last but not least, of Thule or Norway, as far north as the Arctic Circle. No other single traveller known to history has made such far reaching and important discoveries.[31]

The Eastern Expedition of Alexander the Great (Map 7)

While Pytheas was in the north, Alexander and his massive entourage were somewhere in the east, on an expedition

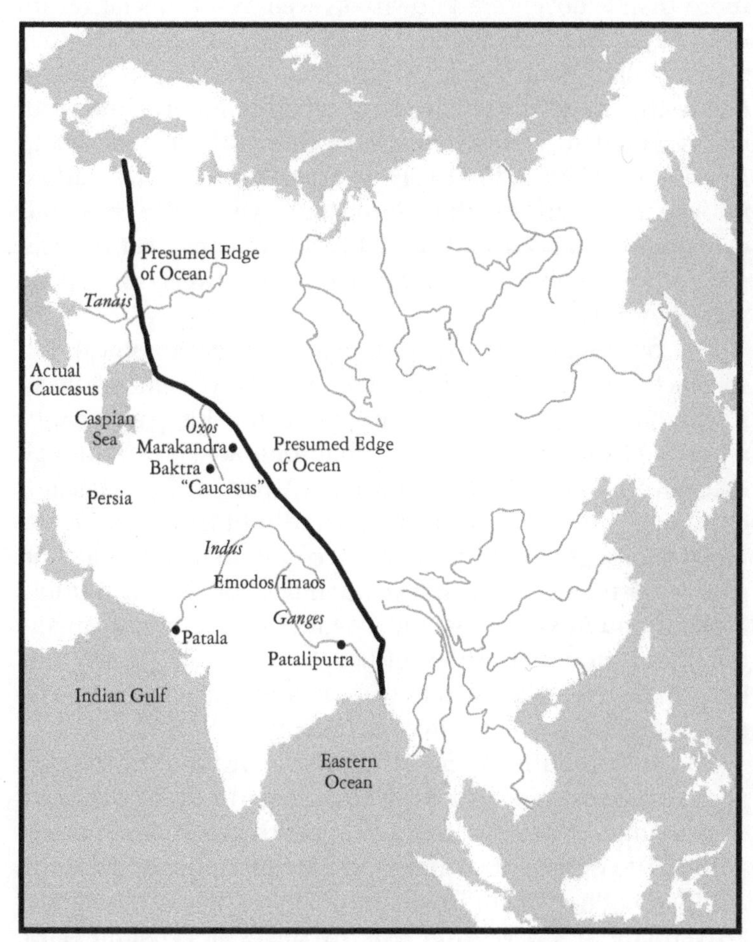

Map 7. Toponyms associated with the eastern expedition of Alexander the Great.

lasting from spring 334 BC until early 323 BC. As with Pytheas, the extant sources are from the first century BC and later; the earliest are Diodoros and Strabo.[32] There were a large number of accounts of the expedition written by the participants—more than a dozen are known—as well as an official record of the journey.[33] Throughout its 11 years, reports and letters were regularly sent back to Macedonia: for example, Krateros of Orestis, one of Alexander's most capable military commanders, reported in a letter to his mother, Aristopatera, about a visit to the Ganges.[34] If the letter is authentic, it is the earliest known mention of that important river. No previous expedition had been so thoroughly documented, yet this was not without problems, and disagreement between the reports was notorious.[35]

The purpose of the expedition was not geographical, but so much new territory was covered that there came to be an inevitable topographical component to the participants' reports.[36] None of these survives, but some are known through extensive excerpts in later authors. Arrian, whose *Anabasis of Alexander* was written in the second century AD and is the most thorough existing Greek account, relied heavily on the eyewitness reports of two members of the expedition, Ptolemy and Aristoboulos, and, needless to say, there are many quotations from both these authors.[37] Others are known only by a single reference, such as Gorgos, a mining engineer, who prospected for gold and silver in India.[38]

The first months were in familiar territory, zig-zagging across Anatolia, and reaching the ancient city of Tarsos in the summer of 333 BC after a year on the road. Then it was around the northeast corner of the Mediterranean and south along the Phoenician coast into Egypt, where Alexander arrived late in 332 BC after lengthy sieges at Tyre and Gaza. The Phoenician cities had generally been on the fringes of Greek knowledge, although Sidon was well known to Homer.[39] It was the first foreign town that Alexander conquered, and he placed a local, Abdalonymos, on the throne, who quickly adopted Greek ways, leaving as his legacy the so-called Alexander sarcophagus, now in

Istanbul.[40] This was essentially the beginning of the establishment of Greek culture among the non-Greeks of the east, the process that epitomizes the Hellenistic world, something that would have its impact on geographical scholarship.

Alexander spent several months in Egypt. It was well known to Greeks, and the expedition added nothing to the geographical understanding of the region. But an event of great significance was the founding of the first Alexandria, formally dedicated on 7 April 331 BC. It was located just west of the Kanobic mouth of the Nile (hence its formal name, usually ignored, Alexandria-next-to-Egypt, as it was outside the Nile delta and thus technically not actually in Egypt). Alexander himself laid out the plan.[41] It was located at the site of an existing village, Rhakotis, and had a fine harbor, near the former island of Pharos (by this time joined to the mainland) where Menelaos had watered his ships.[42] Alexandria was to become the greatest city in the eastern Mediterranean, a role that it fulfilled for the next thousand years.

City foundation, needless to say, was nothing new in the Greek world. The spate of overseas settlement from the eighth to the sixth centuries BC had produced dozens of new towns, extending from the eastern Black Sea to the western Mediterranean. Yet these had been created for economic and demographic reasons, often to relieve pressures in the mother city and, eventually, to exploit a rich hinterland. Alexandria served neither of these purposes: there was no founding city, and the direct economic benefit to the Greek homeland was slight, or not immediately apparent. The population was largely culled from local sources.[43] Its main function was to enhance the reputation of its eponymous founder and to establish Greek culture in an ancient non-Greek environment. Most importantly, Alexandria created a precedent: city foundation of this sort became an obsessive feature of the Hellenistic world, and although some cities prospered and others may have existed only in theory, they caused the rewriting of the map of southern Asia. Alexander alone was credited with 57 foundations,[44] a number of which are still important today, from the first Alexandria itself to Kandahar in Afghanistan, perhaps originally Alexandria in Arachosia, founded in 329 BC.[45]

In the spring of 331 BC, the expedition left Egypt and returned to the Levant, eventually heading northeast into the desert. That summer it reached the Euphrates at the ancient crossing point of Thapsakos, whose location is not identified with certainty, but perhaps was near modern Dibse in Syria.[46] Thapsakos is not mentioned as an active crossing after the time of Alexander, and thus it was soon abandoned, but it remained important to geographers because it became a major node in Eratosthenes' grid plan of the inhabited world, the intersection of a meridian with his main parallel, and demonstrative of how future geographical scholarship depended on data from the expedition of Alexander.[47]

After passing Thapsakos, Alexander was in the region between the two great rivers, the Euphrates and the Tigris, which was called Mesopotamia. Portions of the district had been known to Greeks since the time of Herodotos and his description of Babylon, but the actual term "Mesopotamia" is not cited before the late third century BC, by Eratosthenes.[48] At that time it was still unfamiliar enough that he needed to explain it to his readers, and thus it is probable that the toponym originated in the era of Alexander. If this is the case, it is one of the most familiar geographical terms resulting from the expedition.

Several months were spent in Mesopotamia, and then Alexander headed northeast into Media. By summer 330 BC, pursuing Dareios III of Persia, he was at the Caspian Gates, which would become another topographical nexus for future scholars, eventually seen as the boundary between the southern and northern portions of the *oikoumene*.[49] The toponym was unknown before Alexander's time, and even as late as the Roman period it was believed to be the end of the inhabited world.[50] The gates were a narrow pass, eight miles long and only wide enough for a single wagon, part of the access from the Iranian plateau to the Caspian Sea.[51] The most likely of several passes in this region southeast of Tehran is the Sar-i-Darreh, although determination of the location is hardly resolved. Like Thapsakos, the Caspian Gates became essential to the grid of the *oikoumene*.[52]

Shortly after passing through the Caspian Gates, Alexander came upon the body of Dareios.[53] Nevertheless the expedition continued for several more years, into territory hardly known to the Greek world. The only significant previous report on the region east of the Caspian Sea was by Herodotos, originally from the last days of Cyrus the Great, which was now two centuries old.[54] Yet the memory of Cyrus survived, as a location called Kyroupolis was still an important city.[55] Eventually Alexander turned south into Drangiane and Arachosia (modern eastern Iran and Afghanistan), but the accounts of the expedition during this period place more emphasis on the character of Alexander and the military exploits than topographical detail,[56] and it was only with Seleukid explorations in the following generation that this region became better known.

By the spring of 329 BC Alexander was crossing the Parapamisos range, the modern Hindu Kush, which was given the name "Caucasus."[57] Since passing through the Caspian Gates the expedition had been aware of an extensive east–west mountain range, which could be understood as extending all the way back to the Tauros of southern Anatolia. The western portions—the Tauros and the traditional Caucasus (between the Black Sea and the Caspian)—had long been known, but now it seemed that the range continued far to the east, to the south of the Caspian Sea, where there were high and wooded mountains known as the Tapourians, probably merely the name of the local inhabitants.[58] The mountain range continued east to what was now named the Caucasus. Eventually it would be discovered that it extended even farther, to the Emodos or Imaos (the name is reflected in the modern Himalayas), but this was not realized until the activities of Seleukid envoys of the early third century BC. This belief in an east–west range across the entire eastern half of the *oikoumene*, dividing it in two along a line from the Aegean to India, became a powerful force in the understanding of the inhabited world.[59]

North of the Parapamisos, now called the Caucasus, Alexander and his expedition found themselves in the region known as Baktria, whose main city was Baktra (modern Balkh in Afghanistan). The area had been known to Herodotos as a remote province of the Persian Empire and place of exile, but there were few details.[60] To the north was Sogdiana, even less known, although somewhere in this region Cyrus the Great had been killed. It was all a strange area of large rivers with no outlets and ancient oasis cities, of which the most famous was Marakanda (modern Samarkand), already hundreds of years old, which Alexander made his headquarters from the summer of 329 BC into 327 BC. The rivers were the most conspicuous topographical feature, but they were barely understood. The Oxos (modern Amu Darya), which Alexander crossed near Baktra, was the largest river in central Asia and was believed to drain into the Caspian Sea, thus providing a theoretical connection to the known world.[61] The river actually empties into the Aral Sea (which was hardly known to the Greek world, if at all), although Alexander's local informants told him otherwise.

Farther to the north was another great river, the Iaxartes (modern Syr Darya), flowing northeast of Marakanda. The expedition believed, deliberately or otherwise, that this was the Tanais, the river that, as the modern Don, flows into the northeast corner of the Black Sea and which was considered the boundary between Europe and Asia. Therefore, it was officially announced, Alexander had come to the limits of Europe. The Iaxartes also empties into the Aral Sea, but in antiquity it, too, was believed to flow into the Caspian.[62] Needless to say, the topographical accounts are quite confused in this region, due to a lack of accurate data, whether by autopsy or local information, as well as manipulation by Alexander's recorders.[63]

The one remaining part of the Persian Empire for Alexander to conquer was India. More had been known about India than any of the territory that Alexander had crossed since Mesopotamia; the first account was the report of Skylax of Karyanda, and then there were the comments of Herodotos and Ktesias. At the beginning of 326 BC, Alexander came down into the Indus drainage near modern Peshawar, at the edge of a confusing complex of affluents that were to cause difficulty for topographers for many years. India was seen as a strange and bizarre place, at the end of the world, beyond the stars and the sun.[64] Alexander had heard rumors of another river, the Ganges, larger than the Indus and only a few days away, which drained into the eastern part of the External Ocean.[65] He always tended to believe that he was close to the Ocean, an idea perhaps originating when he was in lower Mesopotamia near the Persian Gulf,[66] and he had expressed similar sentiments while on the Iaxartes. Thus he wished to head for the Ganges and the Ocean beyond, but discontent was building among his entourage and he was forced to turn back to the Hydaspes, a tributary of the Indus, expecting to reach the Ocean by that river. He first thought that the Indus was the Nile, or part of its drainage, largely because of a similarity of flora and fauna, but soon realized from local informants that this was impossible.[67] The Ganges was not to be explored by Greeks for another generation.

Having constructed a fleet, the expedition headed downstream, a slow process that lasted well into 325 BC. Eventually it reached the island of Patala near the mouth of the Indus, where there was a notable city that Alexander used as a base.[68] The delta and the two mouths of the Indus were of interest; this would be the third major delta that Alexander had seen, having previously viewed those of the Nile and the Istros. But the expedition had not seen tides, and most of their ships were wrecked by a strong one, perhaps accompanied by a storm surge. The inexorable tidal cycle, which alternated in leaving ships dry and then underwater, was also a dismal experience that seemed impossible to understand. Ironically Alexander and his men were learning about tides at almost the same time that Pytheas was experiencing them in the North Atlantic.

This part of the External Ocean Alexander called the Indian Gulf, probably the first use of the toponym that would eventually evolve into the Indian Ocean.[69] The members of the expedition did not know exactly where they were, other than far southeast of familiar territory, but the voyage of Skylax of Karyanda, two centuries previously, had told them that it was possible to sail from the mouth of the Indus to Egypt. Alexander, like all educated Greeks of the era, was beholden to the idea of an External Ocean encircling the *oikoumene*, and even though he had probably never previously seen the Ocean, it was natural to assume that known regions could be reached by sailing westerly from Patala.

At the mouth of the Indus, in the summer of 325 BC, Alexander divided his forces into a land group and a sea group; a third portion had already been sent west by an inland route under the command of Krateros.[70] Alexander was to go by land along the coast, and Nearchos was to take the fleet and sail back west.[71] Although Krateros had relatively little difficulty, Alexander's journey was full of hardship, especially in the earlier part, and the accounts emphasize the difficulty in crossing the desert of Gedrosia, one of most desolate in the world. Forced night marches were necessary, supplies ran low, and many were left to die. The primary source (surviving only in later quotations) is Aristoboulos, who had a sharp eye for the nature of the desert, especially its flora, and a vivid ability to describe the terrors of the march, including the incessant climbing of sand dunes, the flash floods, and the rationing of water.[72] It is perhaps the earliest Greek account of the realities of desert travel, although there are few topographical details. In early 324 BC Alexander finally reached Persis.

Nearchos, a Cretan, who had been in charge of the fleet that sailed down the Indus, was placed in command of the sea voyage. He lacked extensive naval experience, and Onesikritos (from Aigina or Astypalaia) was the actual pilot.[73] Both left detailed reports of the cruise, which survive in extensive excerpts, as they were used by Diodoros, Juba II, Strabo, Pliny, Arrian, and others.[74] There are more fragments of Nearchos' work than of Onesikritos', although both authors provided an intensity of detail. Since there are no earlier surviving accounts about a Greek voyage on the Ocean, and the slight remnants of the contemporary *On the Ocean* of Pytheas lack specifics, the two reports are of immense value, although heavily summarized.

Nearchos' version has an epic or heroic quality that can interfere with topographical detail. There are numerous parallels to the *Odyssey*, most notably the description of the meeting with Alexander at the end of the cruise, when Nearchos was dirty and ragged, and believed to be the sole survivor of his expedition.[75] The loss of all his fleet and men was not true, but Nearchos was playing the role of Odysseus. During the voyage, the fleet came across an island where people disappeared after being seduced by a Nereid, and they encountered a sea monster.[76] Even though these are Homeric tales, Nearchos nonetheless explained them: he showed that the sea monster—the earliest detailed report of a Greek experience with a whale—was easily intimidated, and that while the island existed, its dangerously seductive resident was a myth. Strabo astutely pointed out that it was the fear of what they might come across, rather than what they actually encountered, that was the true danger.

Onesikritos' report is more practical and less heroic—as a skilled sailor he knew about the sea—with toponyms (about 20 new ones), and data on the nature of harbors, astronomical phenomena, and the location of water.[77] He is the source of the scant geographical detail from the cruise, described in the form of a *periplous* rather than an epic journey. Yet this coastal route between the Persian Gulf and the mouth of the Indus did not remain in use for long, probably because of the discovery of the monsoon route in the latter second century BC, which cut directly across to southern India.[78] Significantly, all later reports, even from the Roman period, go back to Nearchos and Onesikritos with little if any additional data.[79]

During the last months of his life Alexander prepared for an Arabian expedition. This was a region that was poorly known. Herodotos had had some sense of Arabian customs and the gulfs on either side of the peninsula (the modern Red Sea and Persian Gulf), but the information was vague. He believed that Arabia was the farthest south of all inhabited lands, noted for its rich aromatics and unusual fauna, information probably based on reports reaching Babylonia or Egypt.[80] Skylax of Karyanda had presumably passed along its outer coast in the sixth century BC, but Alexander probably knew nothing about this beyond Herodotos' brief summary.[81] By the latter fifth century BC there were persistent rumors of the great wealth of Arabia, epitomized in the phrase "Fortunate Arabia" ("Arabia Eudaimon").[82] Alexander was well aware of this prosperity, since Arabian frankincense had reached Macedonia in his youth and he was a conspicuous user of the aromatic.[83] His curiosity about the peninsula was consistent enough for later traditions erroneously to suggest that he had actually conquered it. It certainly seemed a reasonable location for his next expedition.

Alexander was probably interested both in the potential wealth of Arabia and in clarifying the route from the Persian Gulf to Egypt: he was still thinking of a circumnavigation of Africa that would allow him to enter the Mediterranean from the west.[84] He first sent out Archias of Pella—who had commanded one of the ships from India—who went down the west side of the Persian Gulf and discovered the island of Tylos or Tyros (modern Bahrain), but did not have the courage to go any farther. This is an odd statement about someone who had not only experienced the entire expedition with Alexander but who had piloted a ship over hundreds of miles of the External Ocean, and one suspects that there is more to the story, especially since Archias is never heard about again.[85]

Next was Androsthenes of Thasos, who had also commanded a ship on the ocean voyage.[86] He left from Teredon, near where the Tigris and Euphrates joined (perhaps around modern Basra) and visited the island of Ikaros (modern Failaka, off the Kuwaiti coast), and then the trading city of Gerrha. He also explored Tylos and nearby Arados (modern Muharraq).[87] Eventually he came to the mouth of the gulf, and continued around the peninsula, but did not reach the Red Sea.[88] He spent enough time on Tylos to observe and record its unusual flora—including the first Greek description of the mangrove—and wrote a report of the cruise, *Sailing Along the Indian [Ocean] Coast,* which was of particular interest to Theophrastos, whose botanical treatises are a major source of its surviving fragments.[89] The treatise was an early example of the effect of Alexander's expedition on scientific research. Eratosthenes, who was probably the last to see an actual copy of Androsthenes' report, found it of geographical value, especially his estimate of the size of the Persian Gulf (a remarkably accurate suggestion that it was slightly smaller than the Black Sea).[90] Nothing more is known about Androsthenes after the publication of his report.

After Androsthenes, there was an obscure Hieron of Soloi.[91] It is not known whether he had been with Alexander on the expedition, and he is remembered only because he was ordered to go all the way around Arabia and into the Red Sea. He turned back, however, probably because of the desolation of the coast and a lack of supplies: any thought that Arabia was a fertile paradise could not be sustained after one left the Persian Gulf.[92] Although the relative chronology of these three explorers is known, their exact dates of travel are not, yet Androsthenes seems to have been on Tylos in the winter (he reported rainfall), perhaps that of 325/4 or 324/3 BC, and Hieron returned in time to file his report with Alexander, probably just before the latter's death in June 323 BC.

There was also Anaxikrates, whom Alexander sent down the Red Sea, and who recorded its length.[93] He may have been the first Greek to visit the frankincense-growing regions at the southwest corner of the peninsula, and he was perhaps commissioned to join up with Hieron, which, needless to say, did not happen. Some of his botanical data may have been used by Theophrastos.[94] The fact remains that the outer Arabian littoral was for many years the only portion of the western and southern coast of the *oikoumene* between the British Isles and the mouth of the Indus that was unknown, if all accounts are to be believed. It is not until the beginning of the first century AD, in Juba's *On Arabia*, that there is evidence that a circumnavigation had taken place, although its date and implementor are not known.[95]

Much of what resulted from Alexander's expedition was raw data, especially in so far as geography was concerned: toponyms and ethnyms, with little overall context. Some of the material was consciously altered to enhance Alexander's achievement. Yet the amount of information was extensive, especially about the regions east of Mesopotamia, and this provided the basis for significant geographical research in the century after Alexander's death.

Manipulation of Topography

Any study of Alexander's contributions to geography faces the opposite problem to that of Pytheas: the Massalian's reputation was denigrated but Alexander's was enhanced, and one way this was done was to manipulate the topography in order to make his achievements seem more impressive. The technique was wisely limited to remote areas that were little known, but it had a significant effect on the topographical history of southern Asia until modern times—an example of how geographical information could be used to deceive rather than inform.

Perhaps the most egregious example was the moving of the Caucasus Mountains. These were the high and rugged range that stretches for hundreds of miles from northwest to southeast between the Black Sea and Caspian, reaching a height of 18,510 feet at the summit of Mt Elbrus. They had been known to the Greek world since the latter sixth century BC and were considered the northern boundary of the Persian Empire of Dareios I.[96] They entered popular consciousness in Aeschylus' *Prometheus Bound* as the highest of mountains, whose summits approached the stars, somewhere in or beyond which Prometheus was imprisoned.[97] Clearly they were a goal for Alexander, but he passed hundreds of miles to their south (in 330–329 BC) as he headed across Media and beyond the Caspian Sea. Eventually he reached the mountains called the Parapamisos, essentially the modern Hindu Kush of north-eastern Afghanistan and northern Pakistan, a range barely known to the Greek world. Evidently the locals, probably at the suggestion of Alexander's people, pointed out a cave in the Parapamisos that was allegedly the one in which Prometheus had been bound, even revealing traces of his chains, perhaps a shrine to a local divinity.[98] An Alexandria was founded nearby,[99] probably to commemorate the site—although the city remained obscure—and the range as a whole was called the Caucasus.[100]

The rationale was to enhance the reputation of Alexander by allowing him to cross the Caucasus, the highest and most rugged mountains in the world. It also gave a mythic quality to his expedition, connecting it not only with Prometheus but also with the Argonauts (Colchis was near the Caucasus). Yet the manipulation soon earned the disdain of professional geographers as well as the new Seleukid state that inherited Alexander's eastern possessions and which depended on accurate information about these remote areas of the world. In the early third century BC the Seleukids sent Patrokles to untangle the gratuitously confused topography of the regions east of the Caspian,[101] and Eratosthenes, later in the century, was outraged at what Alexander's people had done.[102] Strabo pointed out that the Caucasus had been moved 30,000 stadia to the east merely to gratify Alexander.[103] Yet although the professionals knew better, the confusion remained, and even as late as the first century AD there were still two sets of Caucasus Mountains, one the conventional range between the Black Sea and the Caspian, and another to the east near Baktria.[104] In the following century, Ptolemy in his *Geographical Guide* continued the duplication, but had some sense of a problem, citing both the proper Caucasus near Colchis as well as the "so-called Caucasus" in the east.[105]

The regions north and east of the (true) Caucasus also saw topographical adjustments.[106] The Tanais River (modern Don) had long been agreed to be the boundary between Europe and Asia.[107] Alexander never came close to the river, which meant that it could be argued he had never reached the limits of Europe. The territory between the Caucasus and the Tanais was poorly known, as was the northern Caspian Sea, and topography was reworked so that the Caspian was connected to the Maiotic Lake (the modern Sea of Azov, the northern extension of the Black Sea), with the Tanais emptying into the Caspian rather than the Maiotis. In fact, the name "Tanais" was moved to another river, the Iaxartes (modern Syr

Darya).[108] Thus when Alexander reached the Iaxartes in 329 BC, he (calling it the Tanais) could say that he had reached the limits of Europe, having crossed the (misplaced) Caucasus. A certain Polykleitos of Larisa, a member of the expedition, seems to have been the one responsible for this topographical creativity. He provided questionable botanical details to support his case, evidently based on the unexpected lushness of parts of the Caspian coast, which was remindful of the Mediterranean, thus making possible the argument that this region was actually an extremity of Europe.[109] In fact, it was important to Alexander's self-image that he reach the four corners of the *oikoumene*, a desire perhaps based on Ephoros' recent publication of the four ethnic groups that bounded the known world. To Alexander, these limits probably meant the Istros, Nile, Indus and Tanais, with only the last causing difficulty.[110]

Geography was so poorly understood in the northeast corner of the inhabited world that once it started being manipulated all hope of accuracy was lost. The Iaxartes does not flow into the Caspian, but into the Aral Sea. Thus any proper comprehension of the Caspian became totally impossible. In fact the names Caspian and Caucasus were similar enough to be regularly confused.[111] Polykleitos reported that the Caspian had sweet water, but that it was also part of the Maiotis, which is salty (although, granted, with an exceedingly low salinity), demonstrative of how opaque the arguments had become.[112] Moreover, Alexander believed that the Caspian was an inlet of the External Ocean, in part because he wanted easy access to the western Mediterranean, and he repeatedly felt that he was near the Ocean, perhaps wishing to emphasize that he was regularly at the very edge of the inhabited world. That the Caspian was an inlet of the Ocean became the generally prevailing view until the twelfth century.[113] Understanding topography is difficult, and when it becomes deliberately altered it may be almost impossible to comprehend, especially in distant areas. It is no wonder that the Seleukids eventually sent an envoy to this remote region to investigate.

CHAPTER 5

THE LEGACY OF ALEXANDER AND PYTHEAS

The accumulation of data from the journeys of Alexander and Pytheas had its immediate impact on geographical research, both theoretical and topographical, and the following century saw the final movement toward the creation of geography as an academic discipline. The new territories conquered by Alexander became the subject of extensive inquiry. Since he had legally inherited the Persian Empire, the eastern world as far as India was now politically connected to the traditional Greek world, to be explored and analyzed. Even though it was not until the Roman period that anyone returned to the remote regions explored by Pytheas (and the location of Thule was lost), his scientific data were of immense importance in understanding the inhabited world. Moreover, the new Hellenistic governments with their broad territorial extent encouraged the study of topography as a practical need to know about routes, itineraries, and the regions under their control.[1] There were also theoretical developments about the surface of the earth and the extent of the *oikoumene*.

Dikaiarchos and Straton

Dikaiarchos of Messana (modern Messina in Sicily) was one of the handful of known students of Aristotle, but there are few further details about his life other than the implicit date of the late fourth century BC. He wrote a geographical work, perhaps titled *Circuit of the Earth*, surviving in a few fragments that include the first references to Pytheas' journey and the Imaos (Himalaya) mountains, the latter probably from a source close to Alexander.[2] Although the remnants of his treatise are scant, he is remembered for several important innovations in geographical thought.

Dikaiarchos' most important achievement was creating a base parallel across the inhabited world. Following the view of Demokritos, he saw the *oikoumene* as oblong, with its east–west length one-and-one-half times the north–south width. He divided it in two by an east–west line from the Pillars of Herakles through Sardinia, Sicily, the Peloponnesos, Ionia, Karia, Lykia, Pamphylia, Kilikia, and the Tauros Mountains, as far east as the Imaos.[3] The line is remarkably straight except for the inclusion of Sardinia, perhaps an error in the transmission of Dikaiarchos' data. This is the first attempt to make use of the belief that an east–west range of mountains (the Tauros) extended from Anatolia to India, although the southern curve of the Imaos was not yet known. It is also a first step toward plotting points on the surface of the inhabited world. With some modifications, especially by Eratosthenes, this base latitude line became a standard geographical tool for the rest of antiquity.[4] In addition, Dikaiarchos proposed a number of distances within the *oikoumene*, all emanating from the Peloponnesos (where he lived) and probably based on sailors' reports: 10,000 stadia to the Pillars of Herakles, somewhat more to the head of the Adriatic, and 3,000 stadia to the Sicilian Strait (the modern Straits of Messina, at his birthplace).[5] This last figure is the shortest and most accurate of the three; the other two are greatly in error. Nevertheless, Dikaiarchos made the first attempt to calculate long distances in absolute measurements rather than sailing days.

Also of interest is his attempt to determine the heights of mountains. Despite an extant title, *On the Measurement of Mountains in the Peloponnesos*, this was probably merely a chapter in *Circuit of the Earth*. The rugged mountainous terrain of Greece—especially the Peloponnesos—perhaps inspired such thoughts, and high mountains had long been a curiosity, yet earlier comments on their size tended to be more poetic than scientific.[6] Dikaiarchos measured their height using an optical method of triangulation, perhaps employing a mirror.[7] Optics was an emergent discipline in his day, as this was the era of Euclid's work on the topic, which discusses the same technique (although not applying it to mountains).[8] Only one measurement from the Peloponnesos survives, for Kyllene (15 stadia), but others exist for Pelion in Thessaly and Atabyrion on Rhodes, and Mt Olympos in Lykia may be added to the list, although not specifically attributed to Dikaiarchos.[9] The extant measurements are in stadia or Roman paces (probably converted from stadia), and cannot exactly be equated with modern measurements. The theory is sound but there seems to be a consistent error of about 15–20 percent too high. Dikaiarchos' method, refined over the years, lasted throughout antiquity,[10] although estimating the heights of mountains continued to be problematic: Pliny thought that the Alps were 50,000 paces high, or 50 miles.[11]

Dikaiarchos assumed a spherical earth,[12] and may have considered its size. He is perhaps the source for Archimedes' figure of 300,000 stadia for its circumference, which Archimedes made clear was not his own.[13] The number dates from after 309 BC, since Lysimacheia in Thrace, founded that year, is mentioned, so it must be later than Aristotle but before Eratosthenes, and Dikaiarchos seems the obvious choice.[14] This figure seems to have been the result of a rough calculation rather than simply an educated guess.[15] Based on the assumption that Syene (at the First Cataract of the Nile) and Lysimacheia were on the same meridian, the relative position of the constellations overhead suggested that the distance between them was 1/15 of the terrestrial circumference. It was believed that the two points were 20,000 stadia apart, resulting in a circumference of 300,000 stadia.[16] The methodology is extremely rough and is based on erroneous data, mostly notably a failure to realize that the meridians of Syene and Lysimacheia are several degrees apart, since the distance between the two places was based in part on the sailing route from Alexandria to Lysimacheia, by necessity circuitous. But the theory was basically sound, and these figures were certainly in Eratosthenes' mind when he made his detailed calculations somewhat later.

Dikaiarchos also considered the flooding of the Nile, believing that it was due to the influx of the Atlantic somewhere in western Africa, a rather anachronistic theory going back to Euthymenes of Massalia. He further thought that the tides were due to the sun.[17] Neither of these ideas is particularly clearly expressed in the extant sources. Nevertheless, his work on mountain heights, whether a separate treatise or part of *Circuit of the Earth,* and surviving in only three short fragments, seems to have been unique in ancient literature.

Straton of Lampsakos, Theophrastos' successor as head of the Peripatetic school (*c*.286–268 BC), was a natural scientist rather than a geographer, and no work specifically on geography is listed in his bibliography, although a lost treatise, *On Mining Machinery,*[18] suggests a professional interest in the surface of the earth, which may be connected to his curiosity about the formation of the earth and the physical processes of the cosmos. For some time Greeks had been aware that the earth was not static. Volcanic activity would have been the most conspicuous evidence for this, eventually supplemented by Pytheas' report on the strange phenomena at Thule. Xenophanes, Herodotos, and Xanthos of Lydia had all remarked on the changing nature of the earth.[19] Straton considered the question of the causes of these changes, noting that the original outlet of the Mediterranean seemed to have been a channel into the Red Sea (the isthmus between the two was below sea level in many places).[20] He theorized that the numerous rivers flowing into the Black Sea had once caused it to burst out into the Aegean—on the island of Samothrake, just west of the Hellespont, there was a memory of a cataclysmic flood[21]—and this forced the Mediterranean out through the Pillars. Straton was the first to note the current through the Hellespont, and believed that the sea levels had been different in early times. In fact, he thought that seas had once covered the continents. He also felt that the sea bed was uneven, just like land surfaces, that the seas had currents and flows similar to rivers, and that the bed of the Mediterranean had once been dry land. Despite obvious flaws with this theory (the most notable being, as Strabo commented, that the sea bed does not slope like a river), Straton made a serious attempt to reconcile certain visible elements about the physical history of the earth.[22]

Early Ptolemaic Exploration

The Hellenistic states that came into being in the half century after the death of Alexander had profound effects on geography: it was under the patronage of the Ptolemies in Egypt that the actual discipline was created, and a series of Seleukid explorers and envoys examined the new eastern territories in a far more thorough way than had anyone with Alexander.

Alexander's companion and chronicler Ptolemy (I) took up residence in Alexandria-next-to-Egypt, and by 305 BC he had assumed the title of king. For the next 275 years he and his descendants ruled from the city, until the last—Kleopatra VII—fell to Rome in 30 BC. Especially during the time of the earlier Ptolemies, there was intensive support of scholarship, most visibly at the great library that was established early in the third century BC. It became a repository of all Greek literature (and some in foreign languages), including the now lost geographical texts, and served as a research center for scholars. Royal tutors and librarians (often the same person) were the leading intellectual luminaries of the era. Straton was tutor to Ptolemy II, Eratosthenes was both librarian and tutor to Ptolemy IV, and Strabo worked in Alexandria for many years.[23]

Egypt proper had long been known to the Greek world, but the Nile above Meroë, and the coasts of the Red Sea and the Gulf of Aden, were little explored, although Anaxikrates had made a reconnaissance of the Red Sea in the last days of Alexander. These regions were of interest to the Ptolemies for a specific reason: their need for elephants, which had become a requirement of Hellenistic warfare.[24] India, the traditional source of military elephants, could only be reached through Seleukid territory, and thus was not available to the Ptolemies, and the elephants that had come west with Alexander were probably all gone by the accession of Ptolemy II in 282 BC. Yet regular warfare with the Seleukids demanded a supply. Access to the lucrative Arabian aromatic trade was also important. In this context the Ptolemies explored the upper Nile and Red Sea.[25]

The course of the Nile above the First Cataract (at ancient Syene or Elephantine, modern Aswan) had been known to Greeks since the fifth century BC: Herodotos went as far as the cataract, and received reports about points upriver into the Aithiopian territory, a journey allegedly of 56 days partially by boat and partially by foot, until one reached the great city of Meroë, the "Aithiopian capital," located near modern Begrawiyah in Sudan. Two months beyond was the Egyptian Island, occupied by a group of soldiers who had revolted from King Psammetichos (probably the second of that name) in the early sixth century BC, and who had been partially assimilated by the Aithiopians. It was not an island, but appeared as such to those coming from the north, for it was where the two streams of the river, the White and Blue Nile, joined at modern Khartoum. The journey from Meroë to this point was only vaguely known in Herodotos' day.[26]

Seeking to enhance its understanding of this region, the Ptolemaic government began to send explorers south. The exact dates are generally not known, but most were commissioned by Ptolemy II, III, or IV, during the third century BC. The first was Dalion, who went "far beyond" Meroë and reported on the peoples and natural history in an *Aithiopika*.[27] He may have gone to the point in the southern Sudan where the White Nile turns to the west as one goes upstream. He also seems to have included ethnyms from west of the Nile, perhaps learned by hearsay, and he may have been one of the first to believe that the Nile originated in northwest Africa. Herodotos had said that the river flows from the west, which may be nothing more than an awareness of the westward tributaries of the While Nile, but a Mauretanian origin of the river was eventually assumed by Juba II in the latter first century BC.[28]

Other explorers came after Dalion. Aristokreon determined how far the First Cataract was from the Mediterranean (750 miles in Pliny's Latin recension, a reasonable estimate), and also made a detailed report on his journey south of Meroë, although the summary, recorded by Pliny, is difficult to interpret because of the unusual toponyms, which are subject to corruption at all stages of the transmission and have been latinized.[29] Few can be associated with known places. Nevertheless Aristokreon provided the name of the Egyptian Island, Aesar, and located it a mere 17 days south of Meroë (as opposed to Herodotos' two months), certainly a more accurate statement. The toponyms and ethnyms are documented perhaps as far as the region of modern Khartoum, although Aristokreon continued at least eight more days upriver.

Two successive explorers are hardly known. Bion of Soloi wrote in his *Aithiopika* about the culture and institutions of the Meriotic state as well as the topography between the First Cataract and Meroë.[30] Probably the most detailed account of Meroë was by Simonides, who lived in the city for five years during the latter third century BC as a Ptolemaic agent.[31] Unfortunately nothing survives of his writings.

The emphasis of these explorers and diplomats was on ethnography, topography, and natural history, yet they were also very much aware of being far south of the known parts of the world, and it is difficult to believe that they did not record the unusual celestial phenomena and the reality of living in the tropics (Meroë is at 17° north latitude), with the lowering of the northern constellations and the appearance of new stars in the south. A certain Philon may have been sent by Ptolemy II especially to observe these phenomena, as he determined the latitude of Meroë and reported on the strange position of the tropical sun. He was at Meroë long enough to record at least one solstice and one equinox, and thus he became a southern counterpart to Pytheas at the other end of the *oikoumene*.[32] His data were of fundamental importance to Eratosthenes half a century later.

Thus by the latter third century BC the upper Nile drainage, probably well beyond the location of Khartoum, had been added to the *oikoumene*. There was also hearsay knowledge about the swamps and mountains beyond, but there was no further exploration in this direction until the Roman period: clearly the Ptolemaic government had learned all that it needed to know. The upper Nile is no farther south than central India or the frankincense country of southern Arabia, but those regions had not been examined nearly as thoroughly, and Meroë, through the course of the Nile, was more easily connected to the Mediterranean, so it became an important southern anchor in creating the plan of the inhabited world.

There were also Ptolemaic explorers on the Erythraian (modern Red) Sea. As with the journeys up the Nile, the details are scant beyond lists of toponyms. Even though Anaxikrates had previously gone to its mouth, those commissioned by Ptolemy II and III examined it more carefully, with an eye to elephants, the aromatics of southern Arabia, and other items of value. Philon, who was later in Meroë, was stationed on the island of Topazos (probably modern Zabargad) on the west side of the Red Sea, and returned to Alexandria with examples of topaz (periodite) that he presented to Queen Berenike I, wife of Ptolemy I.[33] As a report he wrote *An Account of the Sailing Voyage to Aithiopia*, whose title suggests that sailing down the Red Sea might provide another route of access to the upper Nile. Then Ariston went down the east side of the sea, erecting an altar at Poseideion, the southern end of the Sinai peninsula (modern Ras Muhammed), where the Red Sea begins, perhaps marking the sea as Ptolemaic territory.[34] A summary of Ariston's cruise may be preserved in an extant account of the eastern side of the sea as far as the frankincense territory; its emphasis on aromatics may reveal his purpose.[35]

Satyros explored the western coast on the orders of Ptolemy II, perhaps more than once, looking for elephants,[36] but evidently more information was needed, since Ptolemy III sent

Simmias to the same region for the same purpose.[37] A certain Pythagoras also traveled down the sea, reporting on the available precious stones, mangroves, and fauna in his *On the Erythraian Sea*.[38] A work of a more technical sort was *On Harbors* by Timosthenes of Rhodes, who was a naval commander for Ptolemy II. About 40 fragments exist, and these show that the work covered most of the areas that might have been of interest to Ptolemaic seamen.[39] It may have been a professional manual for sailors, given its title and its inclusion of sailing distances from Alexandria as well as a discussion of the winds. Timosthenes' profession hints that much of the material may have been gathered by autopsy. The work included all of the Mediterranean and Black Sea coasts, from the west in the vicinity of Massalia to Sicily and the Aegean, as well as the Red Sea and the Nile. It seems to have gone beyond an ordinary *periplous*, since Eratosthenes considered it one of his most important sources.[40]

Although these explorers are often little more than names, considering them collectively reveals that by the middle of the third century BC the Ptolemies had covered both sides of the Red Sea from one end to the other, identifying potential elephant hunting grounds, points of access to the Arabian aromatic trade, and other minerals, flora, and fauna that could be economically useful. Many of their reports were summarized in a work titled *On the Erythraian Sea*, written by Agatharchides of Knidos in the middle of the second century BC.[41] *On the Erythraian Sea* is more history than geography, but it included a detailed examination of Ptolemaic interests (primarily elephant hunting) in Upper Egypt and Aithiopia, which led Agatharchides to include geographical and ethnographic detail.

Another element of Ptolemaic activity in this region was the establishment of port cities on the Red Sea that could expedite trade and elephant transport.[42] The most important of these in the early Ptolemaic period was Berenike, about 500 miles south of the head of the sea, at the latitude of the First Cataract of the Nile, named after the same Berenike to whom Philon brought topaz.[43] Originally part of a Ptolemaic plan to create a trade network that avoided the Seleukids, it eventually became one of the major commercial centers of the Roman world, with contacts from the Atlantic to China. Trade items went up the Nile from Alexandria to Koptos (in the Egyptian Thebais, at modern Qift), and then by an arduous overland journey 309 miles to Berenike, which took 12 days and passed through an area of such heat that the caravans traveled by night.[44] Traders of all nationalities came to Berenike—a dozen languages have been documented at the city—which is excellent testimony to the diverse spread of the Hellenistic world and its impact on geography.

Traders at Berenike—and other ports on the Red Sea—knew of a region called the Cinnamon-Bearer (Kinnamomophoroi) Territory. It is not known when this term came into use but it was familiar to Eratosthenes in the latter third century BC.[45] Cinnamon had been known to Greeks since at least the fifth century BC, but its origin was erroneously believed to be in southern Arabia and Somalia. It actually came from far to the east, in southeast Asia, but in the tradition of long-distance trade, had changed hands many times before coming into the Greek horizon on the Somali coast.[46] Its true point of origin was not realized until perhaps the time of Alexander,[47] but the toponym Cinnamon-Bearer Territory continued in use thereafter, located on the Somali coast, which was considered the farthest south inhabited region.[48] As such, it was important in the geographical plan of Eratosthenes, believed to be 3,000 stadia south of the parallel of Meroë.[49] In this way Ptolemaic explorers and traders helped define the limits of the *oikoumene*.[50]

Seleukid Explorers in the Far Northeast

Alexander's companion Seleukos (I) established himself as king in 305 BC and located his capital at the new city of Antioch-on-the-Orontes, founded in 300 BC.[51] In the early years of the Seleukid empire, when its territorial extent was the greatest, Seleukos I sent explorers to examine its farthest areas, in part with the intention of rectifying some of the confusion caused by Alexander's topographers. Demodamas, probably from Miletos, crossed the Silis River, perhaps in response to a local threat.[52] Pliny identified the river as the one "Alexander and his soldiers thought was the Tanais," in other words, the Iaxartes (modern Syr Darya). Demodamas erected altars to Apollo of Didyma, thus honoring his local shrine and claiming the region across the Iaxartes for the Seleukids. The date of his expedition is uncertain, but he was still active professionally as late as the reign of Antiochos I, who came to the throne in 281 BC. Demodamas was Pliny's major source for his own account of this part of the world, but it is impossible accurately to ascertain exactly which specific toponyms or ethnyms in Pliny's narrative came from Demodamas' account. A list of 22 ethnyms, all beyond the Silis, may have been reported by him, as well as two rivers, the Mandragaeus and Caspasus (in Pliny's Latin). Some of the peoples had already been encountered (such as the Massagetai, mentioned by Herodotos),[53] but there are new ones, and some who appear nowhere else.[54] The two rivers are also otherwise unknown, although Caspasus is similar to Caspian, and may be one of the affluents of that sea. Any details of Demodamas' expedition are thus lacking, but he crossed the Iaxartes into central Asia and probably went beyond where Alexander had gone, into modern Kazakhstan, perhaps going farther northeast than anyone else in Greek antiquity.

Patrokles, a high-ranking officer and trusted confidant of Seleukos I, was sent to explore the Caspian Sea and its possible connection to other bodies of water.[55] There is no exact indication of date, although the commission would have been before the king's death in 281 BC. Despite the assumption by Alexander and others that the Caspian was an inlet of the External Ocean, no one had ever been to the place where they allegedly joined.[56] Patrokles, using unpublished documents from the expedition of Alexander, set sail on the Caspian, and reported on the various peoples on either side, as well as the trade route from the Black Sea, whose eastern end was at the outlet of the Kyros River. He eventually reached what he called the "mouth" of the Caspian, 6,000 stadia north of its southern end. He also astutely estimated that the Caspian was almost as large as the Black Sea.[57] There seem to have been no later expeditions on the sea before the Roman period, and thus other details about the Caspian and its surroundings that are not specifically attributed to Patrokles may also be from his report.[58] The distance of 6,000 stadia conforms favorably with the actual length of the sea (about 640 miles), so it seems that Patrokles did cover its entire coast; yet he held to the view that it was connected to the External Ocean and thus one could sail from its mouth to the east coast of India. Either he badly misinterpreted what he saw—perhaps too driven by previous conceptions—or simply agreed with earlier views without truly investigating them. He may have realized that to assume an ocean route to India was politically correct, and that no one was likely to challenge him, as indeed was the case for hundreds of years. There are two major rivers that flow into the north end of the Caspian—the Volga and the Ural—and a curious statement by Pomponius Mela, reinforced by Pliny, describes the Ocean as bursting (*inrumpit*) into the Caspian, whose northern reaches are a channel like a river.[59] This suggests that Patrokles did see the mouths of at least one of these rivers, and assumed it was the connection to the Ocean.[60] Both the Volga and the Ural were known in antiquity, and by the second century AD the Volga (ancient Rha) could be described in detail.[61] Patrokles gathered data about the trade routes east of the Caspian, to India, and believed that the Iaxartes and Oxos (modern Syr Darya and Amu Darya) emptied into the Caspian 80 parasangs (about 320 miles) apart.[62] They actually empty into the Aral Sea, which Patrokles was unlikely to have reached, yet the precise figure suggests a specific report, which he probably obtained by hearsay, confusing the two rivers with ones flowing into the southeastern Caspian. He also made estimates of the size of India, a common activity among Greek scholars of the era. He suggested 15,000 stadia east–west at the north, and 12,000 stadia north–south.[63] His figures are among several from this period, but Eratosthenes did not find his numbers credible.

Patrokles is a difficult personality to assess. Despite Eratosthenes' objections, both he and Strabo tended to find him reliable. By contrast the more mathematically inclined Hipparchos did not, perhaps objecting to his distances.[64] Like Pytheas, Patrokles had the advantage of exploring regions that no one returned to for hundreds of years. His report on the size and coasts of the Caspian, as well as the northeastern trade routes, is the only one before the Roman period, but he was also influenced by the legacy of Alexander, and if he had been commissioned to untangle the topographic errors from that era, he failed to do so.

The Envoys to the Mauryan Court

Greek control over the Indian portions of Alexander's empire did not last long, due to the rise of Chandragupta (Sandrakottos or Andrakottos in Greek).[65] It was said that as a young man he saw Alexander,[66] and he came to power around 317–312 BC, taking control of the Indian parts of Alexander's conquests. Eventually Seleukos I—who probably realized the futility of holding India—recognized Chandragupta's sovereignty.[67] Chandragupta, his son Bindusara, and grandson Ashoka, all played important roles in the spread of Greek geographical knowledge.

Chandragupta established his Mauryan dynasty in northern India late in the fourth century BC, situating his capital at Pataliputra (in Greek, Palimbothra or Palibothra, modern Patna) on the Ganges. Despite the end of any Greek political control over India, he was open to Greek influences, and from the beginning of his reign Greeks lived at the Mauryan court. The first known is Megasthenes, who came to Pataliputra at an uncertain date early in Chandragupta's reign, staying until sometime before the king's death around the end of the century.[68] It is unlikely that Megasthenes was a Seleukid envoy—the empire had not been fully formed yet—but was probably a representative of Sibyrtios, Alexander's satrap of Arachosia and Gedrosia, stationed at Alexandria in Arachosia (modern Kandahar in Afghanistan), and an important figure in the years immediately after Alexander's death.[69] Megasthenes spent several years at the Mauryan court, traveling down the Ganges by an existing royal road and becoming the first Greek to go that far east and to report on the great river and its access to the Ocean.[70]

During his stay, Megasthenes collected material for his *Indika*, a treatise of at least four books, the most thorough work to date on the region.[71] It has not survived, but extensive quotations remain, and it was of great importance to Eratosthenes, Strabo, and Arrian, who provide most of the extant fragments.[72] The *Indika* contained a wide range of ethnographic, cultural, and geographical information, much of it gathered at the court, but also while Megasthenes consorted with a group of wise men who lived under a banyan tree just outside Pataliputra.[73] He gained the confidence of Chandragupta and accompanied him on campaign, and also described Pataliputra in detail.[74] Moreover, he added significantly to the geographical understanding of India, through his own travels and from information received. This included the earliest data on the dimensions of India, more accurate than the later figures of Patrokles.[75] He also had the first conception of the long southward extent of India and listed some ethnic groups in the south, as well as providing a catalogue of dozens of Indian rivers.[76] Greeks were fascinated by the rich fertility of the region, with its two annual harvests and unusual fauna and flora: for example, the *Indika* has the earliest mention of sugar.[77] There were also many marvels, especially anatomically impossible people, best explained through mistranslation and misunderstanding of indigenous customs, coupled with a Greek credulity due to the local exoticism.[78]

Megasthenes also obtained some knowledge of the world beyond India. He was the first to publish a report on Taprobane (modern Sri Lanka). Alexander may have heard of the island, but Megasthenes attemped to locate it and provided some ethnographic details.[79] It was placed seven days south of India, probably reflecting a sailing distance from a point in northern India and thus very much at odds with its actual position only a few miles off the Indian coast, an error that was to cause difficulties when Eratosthenes attempted to create his grid of the *oikoumene*.[80]

There were also the Seres, the Silk People.[81] No details were provided, other than they were extremely long-lived. They were presumably silk traders who came to Pataliputra from somewhere to the northeast, and there is no connection with any specific place, certainly not China, yet contact with them is the first Greek awareness of a world far beyond India. It was only in the Augustan period that further details about the origin of silk came to be known.[82]

Chandragupta was succeeded by his son Bindusara (Amitrochates in Greek) around 300 BC. Daimachos of Plataiai was sent to his court, presumably by the Seleukids.[83] He wrote *On India* in at least two books, which included distances across India and along the route into Baktria, natural history, and certain marvels. Unfortunately his later reputation was poor—Eratosthenes called him an amateur, a view supported by Strabo[84]—and thus his treatise has almost entirely vanished, surviving in only five fragments. It is therefore difficult to determine his contribution to geography, but he may have been the first to connect the position of India to the previously-known parts of the *oikoumene*.

There were others who reported on early Hellenistic India, but they are little more than names.[85] Ashoka, Bindusara's son, who came to the throne about 270 BC and ruled for 35 years, showed his interest in Greek culture by writing decrees in Greek (among other languages) and reaching out to the Greek kingdoms, sending ambassadors to Antiochos II, Ptolemy II, and other monarchs.[86] But there is no record of prominent Greeks at his court, and during his reign contact between the two worlds began to diminish, as both the Seleukids and Mauryans weakened and became more involved with problems closer to home. The early reports, especially that of Megasthenes, were of great use to Greek geographers, particularly Eratosthenes, but after the time of Ashoka no significant new details about India were acquired for more than a hundred years, until the latter second century BC.

The legacy of Pytheas and Alexander was widespread. In the century between the time Pytheas started north and Ashoka died, Greek knowledge of the inhabited world expanded in all directions except the west, where the Atlantic prevented any further exploration, although as early as the fourth century BC there were suggestions of peoples beyond the ocean.[87] Pytheas went far to the north, the Ptolemaic explorers to the south, and those of the Seleukids and others to the east and northeast. The theory of an encircling Ocean had long been taken for granted, but now the idea was believed to have been proven, since by the third century BC it was thought that the inhabited world had been completely circumnavigated except for a small distance in the north.[88] To be sure, this was part absurd exaggeration and part wishful thinking, since this "small distance in the north" was from the mouth of the Baltic to the Caspian, or, in reality, to the east side of India. Obviously no one had made the impossible sail from the Caspian to India, despite Patrokles' assertion that it was feasible. The African coast had presumably been covered, but nothing was really understood about it other than what was near the mouth of the Red Sea and in the northwest. Nevertheless, it was believed that the perimeter of the *oikoumene* was now defined, and this was the last important topographical step before the establishment of a discipline of geography.

CHAPTER 6

ERATOSTHENES AND THE INVENTION OF THE DISCIPLINE OF GEOGRAPHY

Shortly after his accession in 246 BC, Ptolemy III hired Eratosthenes of Kyrene to be librarian at Alexandria, and, eventually, tutor to his son, the future Ptolemy IV.[1] Eratosthenes had been studying and writing in Athens for nearly 20 years, specializing in philosophy and mathematics, and had a reputation as a *philologos*, a learned scholar.[2] Within a few years his academic career would move in a remarkably different direction. As librarian he had access to essentially everything that had been written in Greek, which allowed him to do the research that led to his two profound works on geography: the *Measurement of the Earth* and the *Geography*.

The *Measurement of the Earth* was the earlier work, surviving in only a handful of fragments, mostly preserved by mathematicians.[3] In modern times it was long thought to be a part of the later *Geography*, and it was not until the early twentieth century that it was recognized as a separate treatise, although ancient commentators were well aware of the fact.[4] The complete scope of the work is unknown, but it probably was brief and devoted to the methodology of calculating the size of the earth.

Suggestions for the circumference of the earth had existed since at least the fourth century BC,[5] but these were probably more intuitive than the result of any actual calculations, with the exception of a figure of 300,000 stadia attributed to Dikaiarchos. Eratosthenes refined the rough methodology used to determine that figure, basing his conclusions on the knowledge that Alexandria and Syene (at the First Cataract of the Nile) were on the same meridian (the difference between their meridians is only about 100 miles), and that they were 5,000 stadia apart.[6] This was not an estimate, but the result of a royal survey commissioned by the government (whether for Eratosthenes' benefit is unknown).[7] It was realized from the report of Philon[8] that Syene lay on the summer tropic (actually a few miles to its north, but this made no difference), and thus there was no shadow cast at the summer solstice, something that could be determined by the use of a gnomon (essentially a measuring stick).[9] In Alexandria at the same time a gnomon cast a shadow that formed an angle of one-fiftieth of a circle, which meant that the distance between the two points was one-fiftieth of the earth's circumference.[10] This is a simplified summary of the technique, which relied on a knowledge of ratios, triangles, and a method to measure angles. The result is a circumference of 250,000 stadia, which Eratosthenes adjusted to 252,000, perhaps to provide a number divisible by 60.[11] His figure was not automatically accepted: in the following centuries both Hipparchos and Poseidonios offered their own calculations, resulting in smaller totals.[12] Conversion of Eratosthenes' distance to modern measurements is impossible since the length of the ancient stadion varied, yet it seems remarkably close to an accurate number.[13] Three hundred years later Pliny noted the spectacular nature of Eratosthenes' feat, "expressed by such a subtle argument that one is ashamed not to believe."[14] It was now possible to understand the earth as a whole, and, more importantly, to locate the *oikoumene* properly on its surface, positioning known points, and thus creating an accurate view—whether by a map or merely descriptively—of the world.

Eratosthenes' subsequent work was his *Geography*. This was probably the earliest use of the word (Greek *geographia*), a new term created by him. The treatise was in all likelihood completed before the Roman advance onto the Greek mainland in 218 BC, an event seemingly unknown to Eratosthenes. It survives today in approximately 150 fragments or paraphrases, more than 100 of which were recorded by Strabo, who essentially provided a summary of the work in the first two books of his own *Geography*. The actual treatise probably existed no later than the second century AD: it was familiar to Pliny in the previous century, and thereafter Arrian may have been the last to see one of the rare surviving copies.[15]

The *Geography* was not a lengthy work: only three books are known. There is little evidence that Eratosthenes did much fieldwork beyond visits to Arkadia, Achaia, and Rhodes, and it is probable that the entire treatise was written using the resources of the Alexandria library, supplemented by data received from merchants and seaman in the city.[16] There are about 20 sources cited, from Homer to Eratosthenes' colleague, the mathematician Archimedes, who dedicated his *Method of Mechanical Theorems* to him. Most of Eratosthenes' sources are recent, but the pioneers of geographical thought, such as Homer, Anaximandros, Aeschylus, Hekataios, and Herodotos, were also credited. The work (as summarized by Strabo) is the major source for many of the geographical explorers of the fourth and third centuries BC, such as Pytheas, Androsthenes, Patrokles, Megasthenes, and others. Whether or not a map was included is still debated.[17]

The first book begins with a summary of the history of geography, from Homer to the explorers of the early third century BC. There appears to be a special emphasis on the issues surrounding Homer's role as a geographer, although this may be due to the summarizing of the treatise by Strabo, who was trained as a Homeric scholar and who disagreed with Eratosthenes about the poet's importance. Eratosthenes then continued with a discussion of the physical nature of the earth and its formative processes, citing sources such as Xanthos of Lydia and Straton of Lampsakos. The book closed with a consideration of an increasingly popular genre—fantasy geography—in which mythical locales were created, often for allegorical reasons, and generally placed beyond the limits of the known world. The concept may have originated with Plato's Atlantis, but became common after the explorations of Pytheas and Alexander, as they had penetrated into the strange extremities of the inhabited world.

The second book described the size and shape of the earth. Eratosthenes emphasized that his views depended on a spherical earth upon which all the inhabited portions might not be known, demonstrating that these were still new ideas that were not totally accepted.[18] The arguments in this book are based on the conclusions reached in the *Measurement of the Earth*, and the *Geography* may have included a summary of it, perhaps without the most technical data. The book also dealt with the concept of the *oikoumene*, stressing that it was longer east–west than north–south, and assuming that one could sail west from the Pillars of Herakles and reach India, ideas that had been gestating for some time.[19] There was also an exposition of the five zones, essentially as they had been set forth by Eudoxos of Knidos.[20] But unlike the sparse surviving comments of earlier theories about the *oikoumene*, the actual diction of Eratosthenes is preserved, demonstrating (as he had with the term "geography") his continuing role as an innovator of the language.[21] He marked off an area between two parallels that went around the entire earth, with the *oikoumene* lying between them, and called this shape a *spondylos*, or spindle whorl.[22] In addition, the *oikoumene*, which was an island in the middle of the spindle whorl, was shaped like a chlamys. The word he used, *chlamyoeides*, was new: a chlamys was an outer garment worn by horsemen, most familiar on the riders of the Parthenon frieze. It gave a shape to the inhabited world, a rectangle narrowed at the corners, for it was believed that the inhabited world tapered at its extremities. By using common domestic terminology for his complex concepts, Eratosthenes (who, after all, was better known as a philologist than a geographer) made his cosmic speculations more palatable to non-specialists. Book 2 may also have included some topographical data, especially about the location of Syene, the crucial point in understanding the size of the earth, but most of the topography was reserved for Book 3.[23]

The heart of the *Geography* was Book 3, in which Eratosthenes described the topography of the entire inhabited world, beginning with India and heading west, the reverse of previous and future descriptions: he may have done it this way because of the recent interest in India. He began by establishing his prime parallel and prime meridian (Map 8). The parallel had already been outlined by Dikaiarchos, but Eratosthenes refined and straightened it, extending it from the Pillars of Herakles through the Sicilian Strait (modern Straits of Messina), the southern Peloponnesos, Attika, Rhodes, and the Issic Gulf (the modern Gulf of Alexandretta at the northeastern corner of the Mediterranean).[24] Sardinia, the most anomalous point of Dikaiarchos' parallel, was eliminated, and Eratosthenes' line runs somewhat to the south of that of Dikaiarchos. It was necessary that the parallel go through known points, and thus it was not a straight line, but Eratosthenes was aware of this and the deviation was not great.[25] East of the Gulf of Issos, however, it was somewhat more problematic, defined as "along the entire Tauros" as far as India, passing through two points known to Alexander's expedition, Thapsakos on the Euphrates and the Caspian Gates.[26] The data reported to Eratosthenes told him that the easternmost part of the Tauros (the modern Himalayas) turned to the north, and he realized that this was an error, so he corrected it by making the mountain range extend due east-west. In fact the Himalayas run northwest to southeast, through 10° of latitude, and Eratosthenes' adjustment still resulted in the placement of India too far to the north. Relying on the 15,000 stadia provided by Patrokles for the north–south extent of India, Eratosthenes concluded that the southern end of India (modern Cape Comorin) was on the same parallel as Meroë. In fact that parallel is about 9° north of the one through Cape Comorin, but nonetheless it was an amazing feat to connect two places so far apart, with thousands of miles of unknown territory between them. It was probably the first time such had ever been attempted.

The concept of a meridian—a north–south line across the *oikoumene*—had existed since the fourth century BC, but there

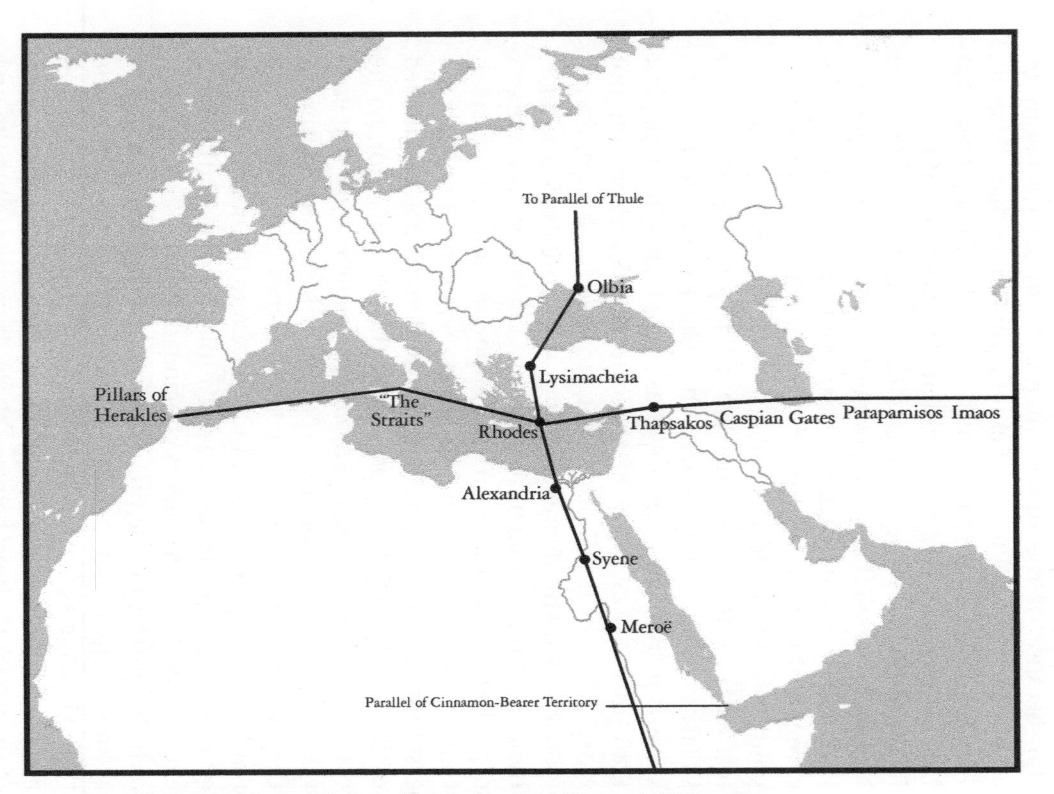

Map 8. The prime parallel and prime meridian of Eratosthenes.

is no evidence of any attempt to plot one before Eratosthenes. Dikaiarchos, however, had come close with his consideration of the relative position of Syene and Lysimacheia. Inspired by the north–south course of the Nile and his previous use of this line to determine the circumference of the earth, Eratosthenes created a prime meridian that was based on the river from Syene to Alexandria.[27] It could be extended to Meroë, also on the Nile and essentially due south of Syene. Farther south was the parallel of the Cinnamon-Bearer Territory, which crossed the Nile. This hypothetical point was said to be 3,000 stadia south of Meroë (somewhere in southern Sudan), but there is no evidence that it was ever plotted on the ground. It was believed to be the southern end of human habitation, because beyond this line it was too hot for people to live. These calculations demonstrated how Eratosthenes' developing grid could determine unknown and inaccessible points on the surface of the earth.

To the north the meridian ran through Rhodes, Lysimacheia, Olbia, and eventually to the parallel of Thule. The line from Alexandria to Lysimacheia, presumably a shipping route, had already been noted by Dikaiarchos, and thus it could be extended to the north. Because of its dependence on a sailing lane from Alexandria, the meridian swung west around the end of Anatolia and then headed northeast across the Black Sea to the Greek settlements on its north shore. It was not easy to plot beyond this point, for even though it had long been known that various Skythians lived to the north, there were no details and no known points or any obvious way to extend the meridian in the manner that it could follow the Nile to the south. Implicit in this, but not actually stated, was that Syene and the parallel of the Cinnamon-Bearer Territory were far closer to the equator than Olbia was to the north pole (and thus the *oikoumene* was in the southern part of the northern hemisphere). For the far north, all that Eratosthenes could do was to run the meridian north from Olbia to the parallel of Thule.

The creation of a prime meridian and a prime parallel—of utmost importance to understanding the location and nature of the inhabited world—allowed Eratosthenes to position the chlamys-shaped *oikoumene* on the surface of the earth. The parallel and meridian crossed on Rhodes, which thus became the nexus of geographical thought. Since the latitude of Syene was known (it was at the summer tropic, the boundary between the temperate and burned zones), and many distances along both lines were available from travelers' reports, or, in the case of Syene-Alexandria, had been measured on the ground, it was now possible to suggest the actual size of the inhabited world: 30,000 stadia north–south by 70,000 east–west. This was a rough figure dependent on many variables: for example, even though the Pillars of Herakles were used as the west end of the prime parallel, it was obvious that land extended to the west both north and south of the Pillars. Eratosthenes, or perhaps Strabo, believed that the "bulge of Europe" added 3,000 stadia to the west.[28] There were also issues about the placement of India; moreover, Thule, whose parallel was now used as the northern limit of the inhabited world, remained enigmatic and its existence even came to be questioned. For various reasons the length of the *oikoumene* was far too long, something that affected Renaissance exploration, since it made the westward route from the Pillars to India appear shorter than it was, an error that actually became worse over time as the far east was added to the *oikoumene*. Nevertheless the perfection of a prime meridian and a prime parallel was another of Eratosthenes' stupendous feats, for it allowed an understanding and sizing of the inhabited world, locating it on the surface of the earth. New parallels and meridians could now be created, even extending to places far from where any Greek had ever been, and the grid of the *oikoumene* could be filled in. As one might expect, much of the extent of the new lines was theoretical or even non-existent, especially the eastern meridians. Few points were plotted at the extremities, such as north of the Caspian Gates or in interior Africa. In theory, however, it was possible to determine the location of any point where a parallel and meridian crossed. Yet the major flaw of the system, as Hipparchos of Nikaia was soon to point out, was that it really depended on information received and known locations, even though some latitudes could be determined by Pytheas' method of the height of the sun at the solstice.

Having placed the inhabited world on the surface of the earth, and determined its size, Eratosthenes then considered its actual nature. He divided it into units that he called *sphragides* (gem or seal stones), a word also used in Egypt to describe a plot of land.[29] Their exact shape is not certain, perhaps rectangles with rounded corners, but this does not easily apply to any of the four sealstones that Eratosthenes proposed. The idea may have occurred to him because India, the first sealstone, was an easily defined unit.[30] This sort of thinking showed the influence of Euclidean geometry, a relatively new discipline in Eratosthenes' day.[31] The idea is important because it looks at geography in terms of land forms, not, as Ephoros had, through ethnicity, yet the concept of sealstones was not truly viable and did not survive as part of geographical theory after Eratosthenes. In fact, Eratosthenes himself began to have reservations, since the sealstones only seemed to work in the east, perhaps because of the large physical units that existed there.

The four sealstones were India, Ariana, Mesopotamia, and Arabia/Egypt/Aithiopia; the amalgamated quality of the last indicates the breakdown of the concept. There was no problem with India and Ariana, but there were serious difficulties with Mesopotamia. The peculiar fourth one, mentioned only once, became virtually impossible.[32] There is no evidence of any sealstones farther west, and the idea did not work at all in the Mediterranean world.

Within the concept of sealstones, Eratosthenes began his discussion of the inhabited world, moving from east to west, and mixing topographical and ethnographic data. About 400 toponyms are known, from the familiar to the still unlocated.[33] The Indian section relied on the explorers of the era immediately after Alexander, such as Patrokles and Megasthenes, and not only considered topography but the routes between India and the Mediterranean;[34] yet the description of the second sealstone, Ariana, is scant.

Mesopotamia was examined more thoroughly but the increasing problems with the sealstone concept tend to burden Eratosthenes' arguments and data.

As was normal for his era, Eratosthenes believed that the Caspian was an inlet of the Ocean.[35] Egypt was described in some detail, and there are comments about coastal Africa as far as the Carthaginian outposts on the Atlantic.[36] The remainder of the *Geography*, from Anatolia across southern Europe to the Iberian peninsula, is sparsely preserved, because these regions, especially west of the Greek peninsula, were better known from other sources by Strabo's time. In fact Strabo, perhaps unfairly, took Eratosthenes to task for his ignorance of the west.[37] Rome is nothing more than an improperly placed toponym between two parallels.[38] Moreover, Eratosthenes, expectedly, relied on Pytheas for the far northwest, whom Strabo believed should be discredited; yet without Eratosthenes (through Strabo), Pytheas and his achievements would be far worse known than they are today. In fact, the most difficult issue in understanding Eratosthenes is that in large part he must be approached through the lens of Strabo, two centuries later, when geographical knowledge had changed significantly, especially in the western Mediterranean, something that Strabo was not generally sensitive to.

The end of the *Geography* is an interesting discussion of ethnicity, and how at the time of Alexander a Hellenocentric orientation had begun to change to a more global point of view with the realization that the Greeks did not have a monopoly on culture and civilization.[39] It was now understood that the *oikoumene* was only a small part of the earth, and the Greek world itself was only a portion of the *oikoumene*. There were advanced cultures—particularly in India—who had not even known about the Greeks. Thus geography was beginning to change traditional views about the peoples of the earth.[40] Eratosthenes gained great fame for his *Measurement of the Earth*, yet his *Geography* was soon forgotten. In 218 BC, as he was finishing the work, the Romans gained their first foothold on the Greek peninsula. Seventy years later Carthage was destroyed. The topographic reality of the inexorable spread of

Roman power resulted in major changes in the Mediterranean world, and within a century of its completion the *Geography* was seen as obsolete. Later scholars with extensive bibliographic knowledge, such as Plutarch or Athenaios, made no mention of it. Yet the fact remains that, however quickly the *Geography* became out of date, Eratosthenes had invented a new academic discipline.

Hipparchos of Nikaia

The most serious early critic of Eratosthenes was Hipparchos, active in the second century BC; the latest date associated with him is 126 BC.[41] He was from Nikaia in northwestern Anatolia but spent most of his academic career on Rhodes. Hipparchos was not a geographer, but a mathematician and astronomer. He wrote a single work of relevance to the history of geography: *Against the "Geography" of Eratosthenes*, which nonetheless is more mathematical than geographical. It survives in 63 fragments, 55 of which are from Strabo's *Geography*, and thus there is the same difficulty in assessing Hipparchos' contribution as there is with Eratosthenes: in fact the two are often tangled together in Strabo's recension, coupled with Strabo's own comments.

Hipparchos' treatise was in three books, the same as the work that he sought to refute.[42] His complaint was that Eratosthenes based his distances and positioning of toponyms on hearsay reports, and he believed, quite reasonably, that the only accurate way to locate places was through mathematics and astronomy, not travelers' accounts.[43] He had mathematical skills that Eratosthenes did not, most notably an ability at what would today be called trigonometry, and developed an astronomical method of positioning places on the surface of the earth.[44] The flaw in his reasoning is not his methodology, but that implementation of such a technique would require a competent observer on the ground at each point, something obviously impossible, at least at that time. Thus Hipparchos was beholden to Eratosthenes' system of information received—especially for remote places—while realizing, as had Eratosthenes, its flaws.

Nevertheless Hipparchos made a number of alterations to Eratosthenes' scheme of the *oikoumene,* most of which were improvements, although some actually made Eratosthenes' data worse. Hipparchos was able to adjust the main parallel somewhat, moving it to the south in the western Mediterranean,[45] and he made other corrections in this region (with the fall of Carthage in 146 BC, the west had become better known to Greeks). He realized that something was wrong with the positioning of India and the Imaos Mountains, but eliminated Eratosthenes' partial correction and reverted to the older, more erroneous data.[46] This may have had a cascading effect elsewhere, since he validated the assumption that Byzantion and Massalia were on the same latitude, when their parallels are actually 2° apart. But he did provide as many astronomically determined locations as possible (although these were few) and outlined the techniques for doing so.[47]

Hipparchos' contribution to geography is not so much his critique of Eratosthenes' methodology as his understanding that an accurate conception of the *oikoumene* was only possible through precise positioning of toponyms and features, and that this depended on astronomy, not travelers' information. It was impossible to apply this theory consistently—something that Hipparchos himself understood—but nonetheless he created the basic theory of mapping.[48] Yet like all such endeavors before the eighteenth century, Hipparchos was hampered by the difficulties in determining longitude, although he realized that observation of eclipses at separate points on the same parallel would provide the information. His statement on the matter, as quoted by Strabo, is a neat summary of the issue:

> It is not possible for anyone—whether an amateur or scholar—to undertake geographical research without determination of heavenly phenomena or the eclipses that have been observed. How far Alexandria-next-to-Egypt is north or south of Babylon—or what the distance is—cannot be determined without an investigation of the latitude. Similarly, the displacement to east or west

cannot accurately be learned without comparing the eclipses of the sun and the moon.[49]

Other Developments in the Second Century BC

A number of other issues affected geographical scholarship in the second century BC but these were essentially tangential to the mainstream. Homeric studies became popular during this era, and it was assumed that Homer had made important contributions to geography, a view already endorsed by Eratosthenes at the beginning of his *Geography*.[50] In fact, the juxtaposition of Homeric studies and topography had long existed, and even served as a way of popularizing geography.[51] Krates of Mallos (in Kilikia), Pergamene ambassador to Rome in the 160s BC, addressed a number of topographical problems in the Homeric poems, believing that the wanderings of Odysseus took place in the External Ocean, emending Homer to this end.[52] He also considered the location of the Aithiopians, something ambiguous in the Homeric poems.[53] These arguments were more philological than topographical, yet of particular interest is that Krates used a globe to illustrate his theories, the first evidence of such a device, and of any attempt to represent visually the entire earth.[54] Whether he actually made one, or just suggested that it could be constructed, remains unclear. It was 10 feet in diameter and showed how limited the *oikoumene* was within the entirety of the world. The globe could also show the parallels and meridians, and how the latter converged at the poles, although whether this was Krates' or Strabo's suggestion is not certain. Krates' topographical views were driven by his Homeric exegesis, and he theorized that the earth consisted of four land masses,[55] separated by streams of the Ocean (thus reconciling the differing Homeric views of the Ocean as both expanse and river), a view that was popular in medieval times but hardly affected ancient geographical scholarship.[56]

Another Homeric scholar of this era was Demetrios of Skepsis, a small town in the Troad (at modern Kuşunlu Tepe, about 20 miles southeast of the site of Troy[57]). Demetrios wrote an exhaustive commentary on the Homeric *Catalogue of Ships*, 30 books long, which was a major source for Strabo. He also considered topographical issues in the tale of the Argonauts.[58] Although it is difficult to imagine how he could have filled 30 books (as long as Ephoros' universal history) on 60 lines in the *Iliad*, he was particularly concerned about the location of Troy.

His younger contemporary, Apollodoros of Athens, who relied heavily on Demetrios' work (Strabo found him derivative), was also interested in the location of places and peoples mentioned in the Homeric poems.[59] Hestiaia of Alexandria, the only woman topographer known from antiquity, was probably also from this era: she, too, wrote on the *Iliad* and seems to have performed fieldwork in an attempt to locate the site of Troy, realizing that the landforms may have changed since early times.[60]

These Homeric scholars of the second century BC, whose topographical efforts are preserved today almost solely in the *Geography* of Strabo, were not truly geographers. Yet Strabo himself was trained as a Homeric scholar, and the relevant passages in his *Geography* are the best surviving example of this type of Hellenistic commentary on Homer.[61] As knowledge of the inhabited world expanded, the Homeric scholars attempted to extend Homer's geographical reach to fit the current reality. They wanted Homer to have been aware of the extremities of the earth, particularly the western Mediterranean and what was beyond the *oikoumene*. Yet they were attempting the impossible: Homer's world was limited to southern Italy and points east. Nevertheless they made important contributions to the study of topography, essentially inventing it as it is still practiced today.

At the end of the second century BC, Artemidoros of Ephesos wrote a *Geographoumena* in 11 books, probably at that date the longest on its topic.[62] Portions survive in an epitome by Marcian of Herakleia, and a papyrus fragment of his text has attached to it a controversial, but perhaps ancient, map of part of Iberia.[63] Artemidoros may have written the first geographical account of western Europe, using contemporary Roman data. Strabo relied on it regularly, and it was one of his most cited authorities, largely for distances, although he did not place Artemidoros in the first rank of geographical scholars. Much of his western data was from autopsy, but Strabo found him less valid in regard to India.[64] Artemidoros calculated the east–west length of the *oikoumene* at 68,645 stadia, a lower figure than usual, probably because he rejected any information from Pytheas.[65]

There was also further interest in the tides. Pytheas had been the first to connect them with the moon, and extensive Greek travels on the Ocean, especially by those with Alexander and the Ptolemaic explorers of the Red Sea, had increased exposure to the phenomenon. Seleukos of Seleukeia (at the head of the Persian Gulf), active in the mid-second century BC, seems to have been the first to write a treatise on the topic.[66] He further developed Pytheas' idea that they were associated with lunar activity, and attempted to establish a full theory of the tides. Tidal activity is still not fully understood, and Seleukos' theory was inevitably incomplete and erroneous. Hipparchos and Poseidonios attempted to verify it but were unable to do so, and Seleukos' efforts came to be virtually forgotten, surviving in only a few oblique references by Strabo. Seleukos was one of the profound but little known personalities of antiquity: he seems to have been the last to support the heliocentric theory of Aristarchos of Samos, even providing a proof.[67]

CHAPTER 7

THE NEW ROMAN WORLD

In 146 BC, Carthage fell to the Romans. One of the immediate effects was the availability of Carthaginian records, since material in the Carthaginian libraries was saved and translated.[1] Moreover, the Carthaginian territory was now accessible, and began to become better known to both Greeks and Romans. Expeditions were sent forth into regions from which they had largely been excluded for many years. To be sure, there had been occasional Greek penetration into Carthaginian territory, most notably on the African coast of the Atlantic, but Carthaginian hostility had kept others out of regions that they considered their own.[2] The fall of Carthage also marks the entry of the Romans into practical geography.

The Roman Republic had existed for more than 300 years but there had been little interest in geographical issues. In addition, much of the early expansionism of the Republic had been into areas already known to Greeks. Although there were no Greek settlements in the earliest Roman territory—from north of the Bay of Naples to the eastern Ligurian coast—the Greek world had been aware of Rome since perhaps the fifth century BC.[3] Even earlier, Greeks had known about the Etruscans, as their settlement of Spina on the upper Adriatic, dating from the sixth century BC, was an outpost within Etruscan territory.[4] The first Roman conquests outside central Italy were among the Greek cities of the south, such as Taras. To the north, the Romans had a relationship with Massalia from the early fourth century BC and used the Massalian treasury at Delphi.[5] The Greeks were familiar with all these regions, and the Romans, even if they were so inclined, could add nothing to general geographical knowledge. When the Romans did become involved in exploration, after the fall of Carthage, their efforts were military in orientation, and generally reflected the political realities of the day.[6]

Polybios and the Romans

It was only after the final defeat of Carthage that Rome moved into areas little known to the Greeks. P. Cornelius Scipio Aemilianus, the conquerer of Carthage, learned about the extent of their territory, and felt the need for a reconnaissance. In his entourage was an explorer of some note, Polybios of Megalopolis, who is remembered today almost solely for his partially extant history but who was best known in his own lifetime for his topographical and scientific research. A few years previously he had traced Hannibal's route across the Alps.[7] Herodotos had been the first to show awareness of the mountains,[8] yet early topographical accounts of south central Europe tended to avoid them, focusing on the Istros to their north. Even Eratosthenes remained vague, although he realized that the Alps were a major range.[9] Polybios may have been the first Greek scholar to cross them, although his (and Hannibal's) route is still disputed today.

After the fall of Carthage, Scipio gave Polybios a fleet and sent him to investigate the Carthaginian world outside the Pillars of Herakles.[10] He went down the African coast "beyond the mountain"—presumably the Atlas—and into the tropics. He visited the old Carthaginian trading post of Kerne, which was in decline (he may have been the last to report on it while it was in existence), and which was abandoned within a century.[11] Eventually he reached the tropics, noting the unusual flora. A number of toponyms and ethnyms preserved by Pliny may have been obtained from Polybios' account but, as was his usual technique, Pliny mixed sources, and it is difficult to determine what Polybios' contribution might have been.[12] Some of the names replicate those reported by Hanno, and Polybios probably had the Carthaginian's report with him: part of his commission may have been to catalogue the ethnic groups of west Africa for the Romans. He may have gone as far as Hanno had, to the mountain known as the Chariot of the Gods (almost certainly Mt Cameroon).[13]

Upon his return Polybios published a treatise, *On the Inhabited Parts of the Earth Under the [Celestial] Equator*.[14] Since the terrestrial equator was invisible—there was no line on the ground marking it—Polybios extrapolated its position from the visible celestial equator, assuming that one lay directly below the other. If he reached Mt Cameroon, he would have been at 4° north latitude, or only 240 miles north of the terrestrial equator. The treatise refuted the long-standing view that the equatorial regions were hot, something Eratosthenes had already questioned, perhaps because of the reports of the Ptolemaic explorers on the upper Nile, who had heard rumors of the mountains of central Africa.[15] Polybios had seen a mountain near the equator—presumably Mt Cameroon—which allowed him to believe that the equatorial regions were in fact at high altitude and therefore temperate. He thus divided the burned zone into two parts, giving a total of six climate zones (rather than the traditional five), and suggested that there might even be a seventh zone, a narrow temperate one at the equator, noting that winds would hit the mountains and produce rain.[16]

Scipio was also interested in the British Isles. He either read something about them at Carthage, perhaps the report of Himilko, or heard about Pytheas, and he and Polybios questioned merchants from the Keltic territory about the region, including ones from Korbilon, the Massalian trading post at the mouth of the Liger (Loire).[17] It is probable that Polybios went to Korbilon to investigate, following the track of Pytheas, since he recorded that he traveled outside Iberia and Galatia (meaning the Atlantic coast of Spain and France), which would mean a voyage on the Atlantic perhaps at least as far as the Liger. But the informants were unhelpful:

> the Massaliotes who were associated with Scipio, when questioned by Scipio about Prettanike, were unable to say anything worth recording, nor were any of the Narbonians or Korbilonians, although these were the most important cities in that region. Pytheas was bold enough to tell these falsehoods.[18]

Thus began the collapse of Pytheas' reputation: the close-mouthed Massalians and their trading partners were not about to reveal the route to the British Isles, or even that they knew anything about them.

Polybios is perhaps one of the more misunderstood scholars from antiquity. Remembered today for his *Histories*—to be sure, an important record of the third and second centuries BC—he saw himself less as an historian than as a new Odysseus: someone whose duty it was to make the remote parts of the *oikoumene* known to the Greeks.[19] Like Ephoros, he included an extensive geographical section in his history.[20] His greatest contributions as a geographer were his exploration of parts of the Alps and his bringing of the Carthaginian researches into the mainstream of Greek awareness, although it must also be remembered that he was instrumental in marginalizing Pytheas. Three centuries after his death, there was still standing in the agora of Megalopolis, Polybios' home town

> a man sculpted on a stele, Polybios son of Lykortas, and an elegiac poem written on it, saying that he wandered over the land and every sea.[21]

There is no evidence that the Romans promptly followed up on the explorations of Polybios, and no Roman presence seems to have been established on the Atlantic coast of Africa. A few communities, such as Lixos, continued as trading centers, and isolated pockets of Carthaginian merchants may have survived into the Imperial period.[22] But the dampening effect of the Massalians meant that Pytheas was relegated to the area of fiction, and the British Isles were ignored until the time of Julius Caesar, a century later.

The Iberian Peninsula

The only European region that interested the Romans immediately after the fall of Carthage was the Iberian peninsula. Polybios had traveled along its Atlantic coast, but this probably had no effect on the Romans, since their concerns were the result of indigenous pressures from the interior. Except for a few trading posts, Greek settlements had not gone beyond the northern coast at Emporion (modern Empúrias), and the Phoenicians and Carthaginians controlled the territory from the Iber (modern Ebro) River south, and also beyond the Pillars of Herakles as far as the Tagus River. Yet by the third century BC both the Romans and Carthaginians were establishing new towns on the Mediterranean coast of the peninsula: Tarraco (Tarrakon in Greek, modern Tarragona) by the Romans, and New Carthage (Nea Karchedon in Greek, modern Cartagena) by the Carthaginians.[23] But little was known about the interior, or the Atlantic coast beyond the Carthaginian settlements south of the mouth of the Tagus River. After the end of the second Punic War (201 BC) the Carthaginians officially abandoned Iberia.

It took the Romans the next two centuries to assert their control over the peninsula, a process that was not completed until the early Augustan period. The result was a thorough knowledge of this rugged territory and its resources. The indigenous stronghold of Numantia, in the north-central part of Iberia, fell in 133 BC to Scipio, the conqueror of Carthage. Polybios may have been present, and wrote an account of the engagement (now lost), but which, one expects, had extensive topographical information.[24] In the far northwest, D. Junius Brutus headed north in 137 BC from the Tagus River and subjugated various indigenous peoples, leaving a report with some topographical details that was used by Strabo.[25] In fact, most of Strabo's Book 3, his account of Iberia, is a composite of Roman data about the peninsula: the first extant topographical survey of this region. Comprehension of Iberia—a long and complex procedure—was the first Roman contribution to geography.

Eudoxos of Kyzikos and the Route to India

After the middle of the third century BC, there had been little contact between the Greek world and India: the close relations between the Seleukids and Mauryans had faded away, and there had been no further use of the sea routes pioneered by Nearchos, Onesikritos, and the explorers of the coasts of Arabia. Eratosthenes had placed India within his grid of the *oikoumene*, and there was trade, yet no one traveled the entire distance between the Red Sea and India. Ptolemaic merchants went only to the mouth of the Red Sea, and Indian ones came no farther west, trading their cargoes somewhere on the southern Arabian or Somali coast.[26]

The situation changed in the latter part of the second century BC when an Indian sailor appeared at the Ptolemaic court and offered to guide a return expedition to India. The tale that follows was preserved solely by Strabo, based on an account in the *On the Ocean* of Poseidonios.[27] The sailor intrigued King Ptolemy VIII, who decided to outfit an expedition to India, and placed in charge a certain Eudoxos, a religious ambassador from Kyzikos in the Propontis who was resident in Alexandria, and who had perhaps already led an expedition up the Nile. Eudoxos made a successful round trip—almost certainly to the rich districts of the west coast of India—and returned with a cargo of precious goods. After Ptolemy VIII died in 116 BC, Eudoxos was sent on a second expedition by Ptolemy's widow, Kleopatra III, and on his return was blown off course and landed somewhere on the west African coast, where he found a shipwreck that had allegedly come from the Carthaginian trading post of Lixos in west Africa.

A falling out with the Ptolemies over his share of the profits led Eudoxos to abandon Egypt and eventually to end up in Gadeira beyond the Pillars of Herakles, where he outfitted his own expedition. His objective was to reach India by going around Africa, avoiding the Ptolemies and their taxes, a variant on the old belief that India could be reached by sailing west from the Pillars.[28] He put together and commanded an elaborate expedition, which was wrecked on the west African coast, and then another, and was never heard from again. This was perhaps around 100 BC; Poseidonios, in Gadeira a few years later, investigated the matter of Eudoxos thoroughly while the locals were still waiting for his return.

The story of Eudoxos, as recounted by Poseidonios and preserved by Strabo, is an epic tale with comic and tragic overtones. Eudoxos was aware of previous circumnavigations of Africa, and believed that Africa was much smaller than it was, which made it seem reasonable that the points he had reached in east and west Africa were not far apart. His effect on geography was twofold: first, he probably discovered one of the Cape Verdes, an island that was "well watered and well wooded but deserted," which he found on his first west African voyage. Nonetheless, no one seems to have returned to the Cape Verdes in classical antiquity.

Second, and Eudoxos' greatest achievement, was the opening of direct trade to India: after his time the route was used regularly. A certain Hippalos, perhaps one of his associates, seems to have established the details.[29] About 20 ships a year made the crossing in the early first century BC, departing from either Mussel Harbor (Myos Hormos) or Berenike on the Red Sea.[30] Mussel Harbor had the advantage of being at the end of a shorter land crossing from the Nile, but Berenike, 1,800 stadia to the south, meant less of a passage on the Red Sea with its adverse winds. Both were used equally. Deteriorating political and economic conditions in the last years of the Ptolemies meant that the route to India was not fully exploited but, after the Romans took over Egypt, business erupted to 120 ships a year, and southern India became well known to the Mediterranean world.[31] Eudoxos also had an effect on Renaissance exploration: Pedro Álvarez Cabral, who discovered Brazil in 1500 while attempting to circumnavigate Africa, was aware of his journeys and was probably inspired by him.

Rome in the Northeast

In the region of the Black Sea, in the early first century BC, Rome became entangled with Mithridates VI of Pontos.

Mithridates' commanders had explored all the northern and eastern coast of the Black Sea, from the Tyras (modern Dniester) River to Colchis.[32] He had Greek scholars at his court, including Metrodoros of Skepsis, who wrote a history that included details about this region and other areas, including India and Arabia and, most interestingly, about the island of Basilia in the Baltic (perhaps modern Gotland or even the Scandinavian coast), possibly information obtained from Pytheas about the amber route.[33] This suggests renewed interest in the Baltic in the early first century BC, perhaps by Mithridates, whose plans were grandiose. If Mithridates knew about the routes to the Baltic from the Black Sea—and since he was an educated man and scholar this was almost a certainty—it is quite possible that he was investigating these connections, and commissioned his court historian to look into the matter.

The third and last war between the king and Rome erupted in 73 BC, and L. Licinius Lucullus pursued him into Armenia, crossing much of the territory that Xenophon had transited more than three centuries previously.[34] Lucullus had perhaps reached the region of Lake Thopitis (modern Lake Van) when his soldiers, appalled at the weather—there was snow on the ground at the time of the autumnal equinox—forced him to turn back. He was in territory that few Greeks and no Romans had seen, although it had long been known that the sources of the Tigris and Euphrates were in this region.[35] Yet Lucullus was soon recalled, and the surviving accounts of his journey—by Plutarch and Appian—have virtually no topographical data about this remote district other than comments on the weather, probably the most severe that any Roman had ever experienced.

By contrast, Lucullus' successor against Mithridates, Cn. Pompeius (Pompey the Great), had an expert chronicler in his entourage, Theophanes of Mytilene, who spent many years with Pompey as his intimate.[36] Although he eulogized Pompey's activities, he was an astute observer, and was well aware of the importance of his role in recording Roman penetration into a region little known. In pursuit of Mithridates in 65 BC, Pompey moved into Armenia, and then beyond, entering the Caucasus and reaching out to the indigenous Albanians and Iberians, with mixed results.[37] He was on the fringes of the known world, and far beyond anywhere Alexander the Great had gone, which may have been part of his motivation. The Iberians, mountain people of the Caucasus, were previously unknown, and Theophanes left a vivid description of them, including how they survived the harsh weather in the mountains:

> There are places in the mountain passes where entire caravans are swallowed up by the snow when major snowstorms occur. They carry staffs against such danger, which they thrust up to the surface in order to breathe clearly and to point themselves out to those approaching, so that they can be helped. They say that hollow lumps of ice form that have good water, which can be drunk by splitting open their covering.[38]

A detailed ethnography and topography from Theophanes' account was preserved by Strabo. The Albanians, on the shores of the Caspian Sea, were slightly known and would have been encountered by Patrokles, but Theophanes was the first to describe them in detail.[39] There were also reports of Amazons in this region but, oddly, they were never seen. Pompey moved on toward the Caspian Sea but soon turned back, allegedly because of the presence of numerous poisonous snakes, perhaps a convenient excuse to extricate himself from a journey that may no longer have had any obvious goal, since Mithridates was far to the west.

Theophanes had an interest in untangling the topographical confusion emanating from the time of Alexander, although he probably did not succeed in this, since there is no evidence that he realized the Caspian was an enclosed sea. There was also a curiosity about the trade routes to India from the Caspian region, and, needless to say, abundant stories about the Amazons.[40] It is not totally clear why Pompey advanced this far: any obvious strategic advantage in the war against Mithridates is uncertain in these remote areas. Going beyond the limits reached by Alexander certainly played a role, and Pompey was interested in exploring the land of Prometheus and Argonaut country. In fact, it was reported that sheepskins were used for gold washing in this region, perhaps an explanation for the Golden Fleece.[41] Pompey may also have been attempting to find out exactly where the coast of the Ocean was, which would bring his expedition to a natural limit: ever since the time of Alexander it had been believed that the Ocean was not far from this area.[42] Reaching the eastern Ocean was certainly a goal of M. Licinius Crassus in his fatal expedition a decade later,[43] and Pompey, whose ambitions were great, may have felt the same way, since the memories of Alexander were a powerful force among many Roman commanders in the east. The Caucasus had been at the fringes of Mediterranean experience since early times, but due to the assiduous efforts of Theophanes, it became the first region in Asia to be explored by the Romans and added to the known *oikoumene*.

Poseidonios

Poseidonios, from Apameia in Syria, was the epitome of the Hellenistic polymath. He wrote on a rich variety of topics, including geography, and validated his research by extensive fieldwork. He was born around 135 BC and lived most of his life on Rhodes, but traveled literally from one end of the Mediterranean to the other, including a lengthy residence in Gadeira, with visits to Massalia, the Keltic territory, north Africa, and the Levant, as well as the traditional Greek world and Italy. He died around 51 BC.[44] His importance to geography is demonstrated by the fact that he is Strabo's third most quoted author, after Homer and Eratosthenes.

His major geographical work was *On the Ocean*, which survives in a lengthy paraphrase by Strabo and numerous other citations.[45] The title reflects Pytheas, a reasonable choice since much of the work concerned the western end of the inhabited world and was the first detailed geographical overview of that region. It was probably written in the early 80s BC after

Poseidonios' extended trip to the west. Strabo believed that it was one of the most important works on geography, equal to those by Eratosthenes and Polybios.[46] His summary, while unusually detailed, was highly selective: it is unlikely that nearly one-quarter of the treatise was about the adventures of Eudoxos of Kyzikos, which is the attention that Strabo gave to the tale, but the story was of particular interest to him for several reasons, including the connections with India, the circumnavigation of Africa, and the geography of the Atlantic coast.

On the Ocean was as much a theoretical and philosophical treatise as one on geography. There was an intensive examination of the terrestrial zones, with a history of zonal theory from Parmenides to Polybios. The recent disputes about the nature of the burned zone were treated in detail, since its limits had been challenged by Polybios. Poseidonios rejected Polybios' division of it into two parts, prefering a more geographically oriented five zones. He also objected to the use of celestial circles to define the zones, since these would vary according to the location of the observer, and he prefered to define the tropics by the direction of shadows, or, following Aristotle, by the habitable and uninhabitable portions of the earth, which would mean five zones. Poseidonios did not object to Polybios' temperate strip along the equator, although he was not particularly impressed by his empirical data, and theorized that the equatorial regions were temperate because the sun moved faster there. This is a point of view peculiar to the modern reader, yet based on the quite reasonable assumption that because the equator represented the longest circumference of the earth, the sun, which circled the earth everywhere in 24 hours, had to move faster over the equator, and thus its heat was less effective. The argument is obscure in Strabo's recension, which often fails to illuminate Poseidonios' more complex points.

The story of Eudoxos follows next in Strabo's summary. Poseidonios' reason for including it was to discuss the nature of the External Ocean, and whether or not it actually circled the *oikoumene*. He learned about Eudoxos while resident in Gadeira in the 90s BC, but in the full text of *On the Ocean* his travels were probably merely a footnote to the larger context of the extent of the Ocean. Poseidonios felt strongly that one could sail to India by going west. This was a view from the fourth century BC,[47] but he emphasized that it was even more feasible than previously thought, for the prevailing east wind and the relatively short distance of 70,000 stadia (a major underestimation) would not make it difficult. This easing of the conditions of sailing westward to India continued: somewhat more than a century later Seneca said it was only a matter of a few days' cruise.[48] These passages were of great influence on Columbus.

In addition to his time in Gadeira, Poseidonios also lived in Massalia, where he gathered information for an extensive ethnography on the Kelts, part of his *Histories*.[49] It is probably the earliest detailed discussion of these peoples: Strabo complained that Eratosthenes knew little about the region.[50] Poseidonios was able to break free from the traditional feigned ignorance of the Massalians about their hinterland, perhaps because he was a resident of their city, thereby adding the territory between the mouth of the Rhone and the English Channel to the known *oikoumene*. In doing so, he wrote the most thorough ethnography to date of the Keltic world, which had been vaguely known to the Greeks since the early days of settlement at Massalia.[51] Poseidonios' description of them is scattered through several sources, in particular extensive summaries by Strabo and Athenaios.[52] It is probable that Diodoros and Julius Caesar also made use of his material.[53]

Poseidonios was also interested in seismology, and seems to have catalogued recent earthquakes, including one in his youth in the Liparaian Islands whose effects he personally investigated some years later.[54] There was also a discussion of how tectonic changes, especially in regard to the level of the earth's surface, affected populations, part of a growing interest in the relationship of geography to demography.

He refined Eratosthenes' method of determining the circumference of the earth, using Rhodes and Alexandria as his terrestrial datum points, and Canopus (α Carinae) as his celestial one, a star not easily visible in the traditional Greek world but conspicuous in Gadeira, which faced south over the sea.[55] Much of his research seems to have centered on the phenomena at Gadeira, whose location provided opportunities not available in the eastern Mediterranean. This included an interest in the tides: he wanted to critique the tidal theory of Seleukos of Seleukeia by comparison with data from Gadeira, theorizing that tidal activity depended on annual, monthly, and daily cycles.[56] Pytheas' rather offhand belief that tides were connected to the moon was developed into a more complex theory, and Poseidonios' examination of the question (seen through Strabo's summary) is one of the fullest extant analyses of the problem, but he had difficulty relating his data to that of Seleukos because of differing conditions between the Atlantic and Seleukos' home region of the Persian Gulf. Nevertheless, there was little doubt after this time that lunar activity was the key to the tides.[57]

There are many other fragments from Poseidonios' works—perhaps not all from *On the Ocean*—that further demonstrate his deep interest in geography, including the depth of the sea and the issue of differing climates at the same latitude. He was also concerned with the exact location of the Pillars of Herakles. Probably because his informants were Gadeirans, he adopted the peculiar view that the Pillars were the bronze stelai in the local temple of Herakles (Melqart), which recorded the expenses of the temple construction, an anomalous view that earned the scorn of Strabo,

> because the Pillars of Herakles should be a memorial of his great achievements, not the expenditures of the Phoenicians.[58]

To Poseidonios, all terrestrial phenomena were part of geography, and thus flora, fauna, and climate all played an essential role, a much more holistic view of the discipline than previously. He was a highly original thinker, and even though Strabo complained that Poseidonios' views were too philosophical, he was an important step forward in understanding geography as an essential element of the total environment. Much of his information was from autopsy during his unsually wide travels, and Strabo, although repeatly objecting to his conclusions, rightly called him the most learned scholar of his era.[59]

The Geography of Julius Caesar

Julius Caesar made important contributions to geography during his time in Gaul (58–51 BC).[60] An educated man, he had read Eratosthenes and probably Poseidonios, who is not mentioned by name in his surviving writings but whose influence is apparent.[61] While in Gaul, Caesar followed the established trade route up the Dubis (Doubs) from the Rhodanos (Rhone) and reached the Rhenos (Rhine), and was the first indubitably to record the existence of that river,[62] although the laconic Massalian traders had probably been that far. The river may have been first mentioned by Poseidonios, but the late source of the fragment makes it uncertain.[63] Caesar learned about its full length (allegedly 6,000 stadia) and some of its tributaries. There was also an awareness of the fertile uplands east of the river, the modern Black Forest. In Gaul proper he catalogued numerous peoples, from the Helvettians in the east (in northwestern Switzerland) to the Namnetians (whose name is reflected in modern Nantes) in the northwest and the Aquitanians in the southwest.[64] Much of his information was from hearsay, or perhaps from Poseidonios, but Caesar's account is the first detailed extant ethnography and topography about the region that is modern France.

He also made two brief trips into Britain, the first recorded since Pytheas. These established "Britannia" as its name rather than Pytheas' more indigenous "Prettanike." Caesar's first crossing was in 55 BC, when he experienced high tides.[65] He was aware that they occurred at the full moon, something that he could have learned from Poseidonios or Pytheas, but was unprepared for the resultant damage to his ships, and thus promptly returned to Gaul. The following year he went back to

Britain for a longer period, and wrote an ethnography and geography of the island, based on his limited autopsy (he was unlikely to have been far beyond the coastal southeast), local information, and probably Pytheas' report. Caesar well knew that merchants went the farthest, and he summoned all whom he could to his headquarters, but was somewhat astonished they could not provide the military intelligence that he sought—perhaps another example of merchants keeping trade secrets to themselves.[66] He was also interested in scientific geography, and used a water clock to determine accurately what had long been suspected: that the summer nights in Britain were shorter than those on the mainland to the south, the first evidence for such a calculation.[67]

Caesar was thus able to provide a new route of access to the External Ocean rather than the traditional one through the Pillars of Herakles. He also made the claim that he had gone beyond the limits of the *oikoumene*, which was one of the reasons that Plutarch coupled his biography with that of Alexander the Great.[68]

In the late summer of 48 BC, the Roman civil war against Pompey brought Caesar to Egypt and its struggling young queen, Kleopatra VII. What happened next is well known, but during those months in Egypt Caesar also indulged his geographical interests. His famous Nile cruise with Kleopatra—remembered in the sources for its opulence—went as far as the frontier of Aithiopia, which would mean at least (and perhaps beyond) Syene and the First Cataract, the center of Eratosthenes' calculations.[69] Caesar, who like so many, was interested in the river's source, probably went farther upriver than had any Roman. It is unlikely that anything new was learned from this voyage, although it was Kleopatra's first reconnaissance of her kingdom and Caesar was probably investigating whether there was anything of interest to Rome in these regions. Furthermore, he had reached the opposite end of the inhabited world from Britain, and thus was able to enhance his reputation. Well read and astute geographically, Caesar was able to extend Roman knowledge both north and south.

Beyond the Known *Oikoumene*

It was believed that the regions north and south of the *oikoumene* were uninhabitable because of extreme cold or heat. To the north, the farthest known peoples were the Roxolanians, and some unspecified populations north of the British Isles, perhaps Strabo's oblique reference to Thule.[70] The Roxolanians lived beyond the Black Sea, between the Borysthenes (modern Dnieper) and Tanais (modern Don) rivers, in other words somewhere in eastern Ukraine, or perhaps farther north. They made war against Mithridates VI,[71] and may have been known to Eratosthenes. They are consistently described as the most northern of all peoples, beyond whom, according to Strabo, it "immediately becomes uninhabitable because of the cold." Despite some inconsistencies—it was known that the British Isles were farther north—there was nevertheless a prevalent theory that one eventually reached a point beyond which it was impossible to live.

The same theory applied to the southern edges of the *oikoumene*, since the parallel of the Cinnamon-Bearer Territory was considered the limit of human habitation.[72] The information was not as vague here as in the north: this parallel was said to be 3,000 stadia south of Meroë, and only Taprobane was believed to be at the same latitude, so a southern limit could easily be defined. Moreover, the parallel marked the line between the temperate and burned zones, and therefore by definition was the end of habitation. Polybios' theory that the equatorial regions were at high altitude does not seem to have created a conflict.

Thus the north and south boundaries of the *oikoumene* could be determined, even if there were some problems of detail. East and west were another matter. The temperate zone in which the *oikoumene* was situated encircled the earth, as Eratosthenes had demonstrated with his spindle whorl, and the east–west extent of the inhabited earth (said to be about 70,000 stadia) was only approximately one-third of the circumference of the globe.[73] This meant that it was more than twice this distance from the Pillars of Herakles around to India (180,000 stadia at the equator, less in the temperate zone). Since the fourth century BC it had been realized that one could theoretically sail around the world from one end of the *oikoumene* to the other, and Eratosthenes noted that

> it is possible that in this same temperate zone there are two inhabited worlds, or more, especially near the circle of Athens that is drawn through the Atlantic Ocean.[74]

Thus anywhere in the temperate zone there could be habitation, yet Eratosthenes also pointed out that this was not the concern of the geographer,

> for the geographer attempts to speak about the known parts of the inhabited world. He omits the unknown parts, as well as that which is outside of it.[75]

Despite this limiting definition of the discipline of geography, there was considerable speculation regarding what lay between the Pillars of Herakles and India. The circumstances were not the same at the two extremities of the *oikoumene*. In the west, the External Ocean had been encountered everywhere, from equatorial west Africa to the British Isles and beyond, so anything west of the known inhabited world had to be in the Ocean. To the east, matters were somewhat more inscrutable, as the External Ocean had only been encountered in two places: on the east coast of India around the mouth of the Ganges, and (in theory) at the mouth of the Caspian. Despite optimistic assumptions of its existence, no one had actually seen the Ocean between those two points, leaving open the possibility that there were unknown parts of the *oikoumene* in the east and northeast.

West of the *oikoumene*, there were repeated reports of islands in the Atlantic, probably fishermen's knowledge of the Madeiras and (less likely) the Azores. Much of the speculation about a world in or beyond the Atlantic was more philosophical than topographical: even as Roman power expanded it was realized that there still might be people living in the remote extremities of the world who did not know about Rome and who were inaccessible.[76] Yet it was also believed that at some future time the Ocean would not be a limit and New Worlds would be discovered, a phrase of great significance to early modern explorers. Thule would no longer be the farthest known land.[77] Yet all specific suggestions of any world west of the Ocean were allegorial, such as the Island of Kronos of Plutarch or the paradise allegedly encountered by Lucian 80 days west of the Pillars of Herakles.[78] The numerous reports of Greek or Roman finds in the New World are either misinterpretations, wishful thinking, or legitimate artifacts that were brought unknowingly to the Americas in early modern times.[79] The only possibly genuine case seems to be the discovery of Roman amphoras in Guanabara Bay (at Río de Janeiro), but, predictably, they were lost before being fully analyzed.[80] Even if these were remains from a Roman ship badly off course—and the sail from west Africa to South America is not difficult, as Cabral learned in 1500—there is no evidence that any news of a world beyond the Atlantic made it back to the Mediterranean.

To the east the situation was different. The Eastern Ocean, with a few exceptions, was elusive, always just out of sight, and thus the possibility existed of unknown regions, especially to the northeast. Demodamas had crossed the Iaxartes and had found land and peoples, but no Ocean. The silk traders at Pataliputra whom Megasthenes saw were also a hint of a farther world, yet nothing is known about them. Other commodities in the Indian trade came from beyond India, most notably cinnamon, which was from southeast Asia and China, but its actual point of origin was unknown to the Mediterranean world.[81] Ginger was also of Chinese origin.[82] These products created some awareness of a world farther east (or north) than India, especially after Greeks were stationed in Pataliputra.

Eventually it came to be known that somewhere beyond India was a place called "Thina." This of course is China, and the name first appears in Greek literature in the mid-first century AD:

> Beyond this land, and at what is already its northernmost point, at some place where the sea comes to an end at the outside, lies a very large inland city called Thina, from which wool, yarn, and the cloth of the Seres are carried by land to Barygaza via Baktra, and to Limyrike via the Ganges River. It is not easy to go to this Thina, for rarely does anyone come from it, not many. The place lies exactly under the Little Bear, and it is said to be contiguous with the parts of the Pontos and the Caspian Sea, where they turn off, and near where the Maiotic Lake—which is parallel—empties into the Ocean.[83]

The passage has uncertainties and ambiguities, but is perhaps the most important topographical statement in Greek literature about the far northeastern part of the world. The trade knowledge that it reflects probably goes back to Mauryan times, since there would be no other reason for one of the routes to have gone down the Ganges. Two routes are described: both go from Thina via the Silk Road to Baktria, and across the mountains into the upper Indus drainage. Then they diverge, with one going straight south to Barygoza, or Bargosa, a major trading emporium in northwest India, probably the place that Eudoxos of Kyzikos visited in the late second century BC. The other heads down the Ganges (this is probably the way the silk traders known to Megasthenes came to Pataliputra) and then around the peninsula to Limyrike, the Malabar coast of southwest India. From Barygoza or Limyrike the goods would go by ship to the Red Sea. There is also a suggestion of a third route directly to the Caspian and Black Sea. Also significant is the assumption that the Caspian—necessarily assumed to be part of the Ocean—might have been adjacent to Thina/China, thereby creating a vast province for the northeastern part of the world that adjoined what was known to Greeks and Romans, closing up any gaps east of the Caspian, and indeed east of the Maiotic Lake. Moreover, whoever reported on these routes did not seem aware of any connection between Thina (a place) and the Seres (its inhabitants), and also located Thina far to the north, at the latitude of the Arctic Circle. The topographical comments are confused, thanks in part to Alexander the Great, and also because the author of the text knew little about the region. Yet there is a concept of extreme remoteness, with "the sea comes to an end at the outside" suggesting the very corner of the *oikoumene*. How early the Silk Road brought its eponymous product to the west is unknown, but one of the first references to Chinese silk in the Mediterranean is in a description of the wardrobe of Kleopatra VII.[84]

In what way Greeks and Romans digested these details about the far northeast remains problematic: Chinese and Mediterranean traders had probably plied the Silk Road for generations, and in the second century BC Chinese came as far west as Baktria, but trade items, not topographical knowledge, were exchanged.[85]

CHAPTER 8

GEOGRAPHY IN THE AUGUSTAN PERIOD

On 10 August 30 BC, Kleopatra VII committed suicide, and after a brief and theoretical rule by her son, Kaisarion, the last Hellenistic dynasty was conquered by Rome, and Egypt became a province. The Seleukids had come to an end 30 years earlier, through the efforts of Pompey the Great and under the authority of the same legislation that had taken him to the Caucasus. Rome now controlled the entire Mediterranean littoral with the exception of northwest Africa and parts of the Levant between Syria and Egypt, which were under the rule of indigenous petty dynasts such as Herod the Great in Judaea. The victor over Kleopatra—Octavian—emerged as Augustus in 27 BC, the First Citizen, or, in modern diction, emperor, and ruled for the next 41 years, creating a dynasty that would last for another half a century, establishing the Roman empire that continued until late antiquity. The extensive Roman territory—from the English Channel to Upper Egypt and interior Anatolia—was highly varied, both ethnically and politically. One of Augustus' main goals was to insure a peaceful empire, and while his primary efforts were military and cultural, new geographical information was steadily gathered.

Egypt and Aithiopia

Egypt was well known. Roman commanders engaged the Aithiopians in the 20s BC, and the geographer Strabo participated in a peaceful journey up the Nile as far as the First Cataract, which he exhaustively documented.[1] These expeditions added detail to the toponymic map of the Nile and to understanding of Aithiopian culture, but at this time there was no further exploration on the river beyond what the Ptolemaic explorers had achieved 250 years previously.

Juba II of Mauretania

West of the Roman province of Africa (modern Tunisia) was a large unorganized territory known as Mauretania (modern Algeria and Morocco). Since the fall of Carthage it had been ruled by indigenous kings, but the last of these had died in 31 BC and thus one of Augustus' priorities was to create a strong pro-Roman government in the region, in part because there were already Roman merchants there. The area was not seen as being ready for direct Roman control but Augustus had a solid pair of candidates as rulers. Juba II had dynastic connections to Mauretania and had been living in Rome since his father, Juba I of Numidia, had been killed in the civil war against Julius Caesar. His wife, Kleopatra Selene, had an even more distinguished lineage as the sole surviving child of Kleopatra VII and Marcus Antonius (Mark Antony), and thus was the only living member of the Ptolemaic dynasty. In 25 BC, the royal couple established themselves at their capital of Iol, renamed Caesarea (modern Cherchel in Algeria).

Juba was a scholar and explorer of note.[2] He had grown up in Rome and had already published a history and other works. When he arrived in Mauretania he immediately began an intensive examination of his kingdom. Kleopatra Selene, with her access to the remnants of the Ptolemaic court, helped implement his research. Their territories were on the geographical and cultural margins of the Greco-Roman world, part of the Carthaginian sphere that had been essentially off limits until 146 BC, and which had received little study since the explorations of Polybios. Juba's goal as a scholar was to learn as much as possible about his kingdom.

A primary interest was the source of the Nile. Since the early third century BC it had been suggested that this might be in northwest Africa, and Juba's explorers mapped out its theoretical route from the Atlas mountains across the Sahara to the known course of the river in Aithiopia.[3] This connected the great rivers of northwest Africa through a series of oases and watercourses to the Nile, and also joined Juba's kingdom with Kleopatra Selene's ancestral territories, a fine example of politically correct geographical theory. Anyone today seeing the river gorges of modern Morocco, such as the Ziz, can appreciate the idea that one or more of these was connected to the Nile. The idea also depended on a point of view popular at the time: that the sources of all major rivers were north of their mouths,[4] a view that was valid for most of the known ones. The Nile would be a conspicuous exception unless it originated in Mauretania. Moreover, there was the assumption that the river ran underground for many miles, and the knowledge that known rivers in west Africa had characteristics similar to the Nile, especially in terms of fauna, a hydrological theory originally advanced by Onesikritos in comparing the Indus and Nile.[5] Even though theorizing a river course with lengthy underground sections may be self-defeating, this became the common view of the route of the upper Nile for many years.[6] The central African mountains where the Nile actually originated had been known since the fourth century BC, but in a shadowy fashion, yet even after they were specifically located around AD 100,[7] Juba's theory continued to be accepted, and it was not until John Hanning Speke discovered the actual central African source of the river in 1862 that thoughts of a northwest African origin were abandoned.[8] Regardless of the validity of his point of view, Juba added much new information about the rivers of northwest Africa, the Sahara oases, and the actual Nile itself in Aithiopia.

Juba's other major discovery was the Canary Islands. They were visible from the coast, and probably had been known since Carthaginian times, but details are lacking, and there was no complete survey of the group until that of the king. He allegedly named them after the dogs found on one island:

the extant name, Canaria, is from Pliny's Latin, but Juba wrote in Greek, and his name may have been Kynika. This approximates that of an ethnic group in west Africa, the Kynetians, so the name may have actually had an ethnic origin. Juba catalogued all the islands in the group, giving them descriptive names, none of which has survived into modern times except Canaria.[9]

Juba published his research in a treatise titled *Libyka* (using the ancient name for the entire continent), which survives today in 36 fragments, most of which are in the *Natural History* of Pliny. Pliny's use of Latin can cause problems in interpretation, especially understanding the many toponyms. There is also a fair amount of unattributed material in Book 17 of Strabo's *Geography*: Strabo and Juba probably knew one other, and the former in all likelihood received unpublished information. Many of the fragments of *Libyka* are about natural history rather than geography, including the most complete extant account of the north African elephant, now extinct.

Libyka also contained a description of the coast of east Africa, which lists peoples from the Mossylitian Promontory (almost certainly Cape Guardafui in Somalia, the easternmost point of the continent) down the coast.[10] The ethnyms are for the most part otherwise unknown except for the Zanganai, 1875 miles south of the Mossylitian Promontory, remindful of Zanzibar, which is in the proper area. The original source of the material is perhaps Agatharchides of Knidos—although this is not certain—but Juba's report was probably the most detailed to date about the farther areas of the east African coast.

In late 2 BC, or early the following year, Augustus' grandson, Gaius Caesar, embarked on a major expedition to Arabia.[11] Juba was asked to join as one of Gaius' scholarly advisors. Gaius and his entourage sailed to Gaza on the southern Levantine coast, moved inland to the Nabataean capital of Petra, and eventually went as far as Aila (at modern Aqaba). He may not have entered the Arabian peninsula itself, but the political context of the journey—to stabilize the region in the uncertain period after the death of Herod the Great—was accomplished, and much was indirectly learned about the peninsula, especially from sources in Gaza and Petra.

Arabia had long been a mysterious and shadowy region, whose reputation was enhanced by being the source of rich aromatics.[12] The first attempts to understand it topographically dated from the period of Alexander, yet his explorers failed to cover the entire coast of the peninsula. In the 20s BC, Aelius Gallus, Prefect of Egypt, made an expedition into its western regions, motivated by a desire on the part of Augustus to learn more about the area and to gain direct access to the aromatic-producing Sabaeans, thereby eliminating the Nabataean middlemen at Gaza and Petra.[13] Yet Gallus' journey was a disaster due to his inability to understand the rugged and barren terrain and his entanglement in a virtual civil war among the Nabataeans. He landed at White Village (Leuke Kome), near the northern end of the Red Sea on the west coast of Arabia,[14] and spent several months in a difficult and futile journey toward the aromatic territory in the south, eventually retreating (with great loss of life) across the Red Sea to Mussel Harbor. Strabo, who was on Gallus' general staff but did not participate in the expedition, nevertheless reported on it thoroughly, making some attempt to attribute its failure to the meddling of the Nabataean royal minister, Syllaios. Nevertheless Strabo was hardly able to hide the fact that Gallus was actually to blame. Despite its inability to achieve its political or economic goals, the journey added details about the topography of western Arabia.

At some time after the era of Alexander the remaining coast of Arabia—that from the mouth of the Persian Gulf to the Red Sea—had been explored. The first report on the coast of the entire peninsula is by Juba, using material that he gathered while on Gaius Caesar's expedition. The account, in the style of a *periplous*, was preserved by Pliny, and begins at the town of Charax, a trading center founded by Alexander at the head of the Persian Gulf.[15] Juba probably visited here, and it was the home of another geographical scholar, Isidoros, who was also one of Gaius' advisors.[16] Isidoros is remembered for his *Parthian Stations*, a list of places and distances from Zeugma, a crossing point on the Euphrates, to Alexandria in Arachosia, which is the earliest extant detailed account of the route east. He also wrote about the Arabian peninsula, and may have been Juba's source for his own *periplous* around the coast, perhaps ultimately based on sailors' data from Charax.[17]

The result of Juba's research on the Arabian peninsula was his *On Arabia*, published in the early first century AD. It was the most complete account of the peninsula to date, although he probably never visited it and relied on material from merchants and sailors at Gaza, Petra, and Charax. The caravan route from the aromatic territory in the southwest to Gaza is outlined, with a detailed report on the process of harvesting frankincense and myrrh.[18] There is also a description of the sailing route east to India, but here Juba had to rely largely on the reports of Nearchos and Onesikritos, with minimal updating.[19]

On Arabia and *Libyka* can be considered as a unit, collectively describing the entire southern coast of the *oikoumene* from the Pillars of Herakles to India, thus closing the long-existing gap on the Arabian coast. Juba, remembering Eudoxos of Kyzikos, may also have wondered about the feasibility of reaching India and its wealth by going around Africa: he even believed that the Atlantic Ocean began at the Mossylitian Promontory on the Somali coast and that it was an easy sail from there to Gadeira.[20] Augustus also was interested in the Indian trade, and made intensive use of the Red Sea/ Indian sailing route discovered by Eudoxos, which had never been fully exploited by the Ptolemies. In the first decade of the Augustan period, this trade was already several times what it had been in the last years of the Ptolemies.[21] Yet Juba never implemented his own trade route to India, and it would have been politically incorrect to do so, but 1,500 years later Vasco da Gama proved its viability when he traveled to India from Portugal and brought eastern trade around Africa to Europe.

Northern Europe

Greek knowledge of northern Europe was limited, with little data from beyond the Alps (which themselves had hardly been penetrated) or the Istros (modern Danube), except along the coast of the Black Sea. Pytheas' information on the Baltic was largely forgotten by Roman times, and the amber routes across the continent provided hardly any topographical detail. Only the northwest—the Gallic territory—was understood, through the efforts of Poseidonios and Caesar, but east of the Rhenos (modern Rhine) and north of the Istros there remained into the Augustan period a large number of unexplored and immense forests. Their existence had been known since the time of Eratosthenes but they had not been penetrated.[22] Yet the indigenous peoples of these northern regions could be a threat to Rome, as had happened in the early fourth century BC when Rome had been destroyed by the Gauls. Augustus believed that the Alps and the regions to their north had to be secured.[23] His stepsons, Tiberius (the future emperor) and Drusus, moved into the mountains in 16 BC, eventually reaching the body of water now known as Lake Constance or the Bodensee. Strabo, the earliest source, described it in detail but failed to provide its name, which suggests that Tiberius did not have one in his report. Several are documented by later authors: Venetus, Acronus, or Brigantinus—this last surviving in the modern lakeside Austrian town of Bregenz—probably representing different names attached to different parts by different peoples.[24] It was known that the Rhenos (modern Rhine) originated in this region—it actually flows through the lake—and Tiberius learned that he was near the source of the Istros and visited it. It was allegedly one day's travel from the lake, which would be more accurate about the river itself than its source, today defined where several tributaries come together at Donaueschingen to create the Danube, another 15 miles up the river from where Tiberius probably approached it. Nevertheless it is clear that he had the distinction of discovering the headwaters of the longest river in Europe.

The upper part of the Istros was called the Danuvius, which eventually became the name for the entire river.[25] It is not uncommon for long rivers to have different names along their course, as rarely does one person know its entire length.[26] The name is first documented by Caesar (as Danubus) and

Diodoros (as Danoubios); neither knew that it was the same as the Istros, and Diodoros thought it flowed into the Ocean, perhaps a confusion with the Rhine or Albis (modern Elbe).[27] Strabo was the first to connect the Danube (which he called the Danouios) and the Istros,[28] with the change of name occurring at "the cataracts," the modern Iron Gates on the Romanian/Serbian border. Strabo was writing only a few years after Tiberius' campaign, and was the earliest to report on the entire length of the river, with the middle portions defined by Tiberius and others during the last years of Augustus. These expeditions also determined the means of access from the head of the Adriatic (at Tergeste, modern Trieste) to the river, as well as details about the river systems of modern Slovenia and Serbia.

The territory north of the Danube and east of the Rhine was also important to Roman frontier policy. Caesar had gone across the Rhine briefly in the 50s BC, constructing a bridge (which he described in detail). He spent 18 days east of the river engaging the Sugambrians, since they were harboring raiders who had come west. Later he crossed again, for similar reasons, building another bridge: both were probably in the vicinity of modern Koblenz.[29] Like his raid into Britain, these incursions provided little topographical detail.

In 12 BC Augustus decided to move across the Rhine and advance the Roman frontier to the Albis (modern Elbe). His stepson Drusus reached the river in 9 BC after a difficult march through marshy country and dense forests totally unlike anything known in the Mediterranean world.[30] Strabo's report that the Rhine and Elbe were 3,000 stadia apart—about twice the maximum distance—demonstrates the circuitous path of the Roman advance. The rivers of Germany became known during Drusus' campaign, such as the Bisourgis or Visurgis (modern Weser), the Loupias or Lupia (modern Lippe), and Salas (modern Saale), near to which Drusus fell from his horse, dying after 30 days.[31] The Vistula is first mentioned half a century later;[32] whether this was information gathered by Drusus or his successors is unknown. The citation may even be a remnant from the days of Pytheas, or traders' information.

Drusus also became the first Roman to sail on the North Sea, at the mouth of the Amasis (modern Ems) River and the island of Byrchanis (probably modern Borkum, the westernmost of the East Frisian Islands). After Drusus' death other Roman commanders, including Augustus himself, spent the next 20 years making expeditions east of the Rhine.[33] There is no evidence that anything was known about the territory east of the Elbe except by hearsay, and the region between it and the Rhine was never secure, although there was extensive trade between Italy and Germany for many years thereafter.[34] In AD 9 the loss of three legions under L. Quinctilius Varus resulted in the eventual Roman abandonment of any interest beyond the Rhine.[35] Because of this, geographical writers such as Strabo, Tacitus, and Ptolemy could not present a coherent view of this region: as Tacitus wrote, the Albis had become no more than a name.[36] Inconsistencies of interpretation lasted until the end of antiquity, including the location of Scandinavia, which may have been known to Pytheas but which remained only an uncertainly placed toponym.[37]

The Far East

Relatively little was added to any understanding of the eastern parts of the world during the Augustan period. Pompey the Great and Theophanes had explored the Caucasus, and Isidoros of Charax was able to produce an itinerary of the route to Arachosia, perhaps taking advantage of the peaceful conditions between Rome and Parthia after 20 BC. The allied king of Kappadokia, Archelaos, the third of the scholarly advisors of Gaius Caesar during his Arabian trip and the most senior of the kings (one of the last survivors of the era of Kleopatra VII and Antony), was Augustus' primary informant on affairs of the east. Like Juba, he was also a scholar, and wrote a work on Alexander the Great that probably updated the route to India, with a view toward Roman policy.[38] He was described as a "chorographer," a word devised by Strabo to mean a topographical scholar,[39] but unfortunately only two fragments survive of Archelaos' treatise, one ethnographic and the other on natural history,[40] so his role in refining geographical understanding remains unknown.

Most of the information used by Archelaos and Roman officials probably came from merchants, since there were no Roman expeditions to the far east. There were two main routes east (and many branches): the traditional Silk Route, toward the north, and the one described in Isidoros' *Parthian Stations*, which was essentially that followed by Alexander. They joined and diverged in Baktria. Frequent mention of the Silk People—as well as Baktrians and Indians—by the Augustan poets demonstrates that the routes to the exotic east had entered popular culture.[41] Vergil, in a long geographical passage in the *Georgics* largely designed to demonstrate the superiority of Italy over the rest of the world, briefly described how it was believed that the Silk People (Seres) produced silk.[42] This is the first actual connection between the people and the product itself: to Megasthenes, Seres was merely an ethnym, which, to be sure, later critics have reasonably assumed to refer to silk producers.[43] Silk was as yet little known: Vergil believed that the raw product was a leaf that was combed into silk ("foliis depectant tenuia Seres"). It is not until the second century AD that there is an actual description of the silk-making process.[44]

These mercantile connections between the Roman world and the east—120 ships a year on the Red Sea/India run in the 20s BC[45]—resulted in embassies from the remote east to the Roman world. Chinese and Romans probably did not come face to face until the second century AD, but Augustus could claim that Indians often came to him, and they had never before been seen by a Roman commander.[46] In 25 BC he received an Indian deputation while he was resident in Tarraco (modern Tarragona) in Iberia, about which little is known.[47] Five years later a delegation came to him at Antioch in Syria, which Nikolaos of Damascus (who may have been present) described in detail.[48] The ambassadors were sent from Poros, who styled himself as ruler of 600 kings (he sent a letter in Greek to Augustus), and who offered friendship and free passage through his territory. He sent numerous gifts, including a man without arms named Hermas, whom Strabo saw at some time. The gifts included exotic fauna of a type probably never before seen in the Mediterranean world. Poros' reason for sending the embassy is unknown beyond the obvious: India and Rome were engaged in extensive trade, and there might have been a value in establishing relations at the highest level. Augustus' response has not survived, and India remained a land of exoticism and myth in the Roman imagination, a subject for poets and imperial posturing.[49]

Marcus Agrippa and his Map

M. Vipsanius Agrippa, the longtime friend and primary advisor of Augustus, traveled over much of the Roman world, especially during the last decade of his life, from 23 to 12 BC. During those years he visited Gaul, the Rhine, and Iberia, and made several trips to the eastern Mediterranean.[50] He had a reputation as a diligent geographer,[51] but little is known about his writings. His primary contribution to geography was the creation of a map of the inhabited world (not merely limited to Roman territory), which was completed by Augustus after Agrippa's death and set up in the Porticus Vipsania in Rome.[52] Agrippa seems to have written a commentary to go along with it, although this is merely assumed from the diction of Pliny, and is by no means certain. Pliny's primary interest in the map (or its commentary) was measurements, and thus there is a lack of detail in terms of exactly what it depicted, and how it was presented. Nevertheless the various citations demonstrate that it covered a wide area: including Britain, Germany, Iberia, the complete Black Sea coast, the Caspian, India, Aithiopia, and Lixos and Hanno's Chariot of the Gods on the African coast of the Atlantic.[53] It also included a peculiar swath from the Danube to the Vistula, 1,200 miles long and 396 wide, which cannot easily be explained, but which seems to represent, in some way, a trade corridor. The citations by Pliny are more frustrating than informative, but making the map involved a prodigious amount of research, incorporating such obscure sources as Pytheas and Hanno: in fact, probably every possible geographical account available in Rome during the latter first century BC. The map may have been based on, and updated from, one started by Julius Caesar just before his death.[54]

It is unfortunate that so little is known about the map and its possible commentary, as the wide range of toponyms preserved in Pliny's sparse references demonstrates that it covered remote areas only slightly known. Information from it probably pervades the geographical part of Pliny's *Natural History* without attribution, especially noticeable with his seemingly irrelevant use of "above" and "below," showing that he was looking at a map on a wall.[55] But nothing has survived from the physical map itself, or the building in which it was located.

Perhaps connected in some way was a source Strabo used several times: the Chorographer (no other name is provided) or a describer of territory, a word that Strabo may have invented.[56] The Chorographer was quoted for Italy and Sicily, and used miles, so was Roman, or at least romanized, but there has been no agreement as to who this writer was.[57] Moreover, Vitruvius, Strabo's older contemporary, wrote about "the sources of rivers painted or written on chorographies of the world," another enigmatic reference (and the first use of "chorographia" in Latin).[58] Whether this refers to Agrippa's map and possible commentary remains unknown: Vitruvius' use of the plural suggests that maps of the world were a common feature in Augustan Rome.

Strabo of Amaseia

The culmination of ancient geographical scholarship is the 17-book *Geography* of Strabo of Amaseia, completed in the 20s AD. Without the survival of this work, almost nothing would be known about either the history of ancient geography through the Augustan period, or the state of knowledge of the discipline at that time: the numerous citations of the *Geography* in the present work demonstrate its importance. It is the only treatise of its type to survive from antiquity.[59] Strabo was born in the 60s BC, a member of the aristocratic elite of the kingdom of Pontos in northern Anatolia. His family had long been in the service of Pontic royalty, especially

Mithridates V and VI, which gave Strabo access to extensive internal information about the history and geography of Anatolia and the Black Sea region during the last two centuries BC. With the collapse of the kingdom of Mithridates VI and the arrival of the Romans, the family moved to Nysa in Karia, an important cultural center, where Strabo associated with the leading intellectuals of the era and was trained primarily as a Homeric scholar, a point of view that was to pervade much of the *Geography*.[60] He then went to Rome and, eventually, Alexandria, where he lived for many years, part of which was on the staff of the prefect Aelius Gallus in the 20s BC. He traveled extensively during his long life; although he never went west of Italy, he was thoroughly acquainted with everywhere else in the Roman and Pontic world. A strong interest in mining and quarrying—discussed more than 100 times in the *Geography* and including a command of the technical vocabulary—suggests a possible professional career. In his later years he became part of the circle around Pythodoris of Pontos, queen of his ancestral region since the end of the first century BC—one of several women of talent and ability profiled in the *Geography*—and lived either in his home town of Amaseia or her nearby capital of Caesarea.[61] The latest datable citation in the *Geography* is the death of Juba II of Mauretania in AD 23 or 24, and presumably Strabo died shortly thereafter.

The *Geography* is a complex treatise, one of the longest surviving from antiquity, covering a wide variety of topics beyond the purely geographical, including Homeric criticism, cultic history, biography, autobiography, and linguistics. Before starting the work, Strabo had written a general history (beginning in the second century BC where Polybios had ended) and an account of Alexander the Great, portions of which ended up in the *Geography*. There also may have been works on Homeric criticism and cultic history, which are themselves reflected in the geographical treatise. This complex origin, and the sheer length of the *Geography*, as well as a gestation period of 50 years, means that it can be inconsistent and difficult to read. The vocabulary is often unique, as is common in Hellenistic scholarly treatises, a difficulty enhanced by Strabo's frequent quotation of lost works whose own diction was unusual. The range of material used is astonishing: simply put, the entire extent of Greek literature from Homer (cited more than 700 times) to his own contemporaries, and several Roman sources (unlike many Greek scholars working in the Roman world, Strabo was fluent in Latin). Were it not for the *Geography* there would be almost nothing known about the geographical writings of Ephoros, Eratosthenes, Hipparchos, Polybios, or Poseidonios. This is not without pitfalls, however, since these authors by and large can only be seen through Strabo's eyes, and that of the environment of the early Roman Empire. In addition, the treatise is a rich collection of citations from lost works by familiar authors such as Aeschylus, Pindar, and Aristotle.

In its 17 books the *Geography* makes a circuit of the *oikoumene*, beginning (in the fashion of Hekataios of Miletos) with the Iberian peninsula. Introductory to this, however, are two books on the history of geography: it is here that the modern reader learns about Pytheas' journey north, Eratosthenes' grid of the *oikoumene*, Hipparchos' criticism of his methods, the strange career of Eudoxos of Kyzikos, and Poseidonios' *On the Ocean*. The 15 remaining books are the geographical account proper, with many historical and cultural digressions. Strabo preserved some of the earliest extant material on Alexander the Great, and many of the surviving fragments of Megasthenes' *Indika*. His attempts to locate the sites of Troy and Nestor's Pylos are early examples of the discipline of topographical research.

A work this long, and completed during half a century of great changes in the Roman world, can at times lose sight of its overall goal, which may actually have changed as Strabo's writing progressed. In fact, there seems to have been a dual purpose, one for Romans and one for Greeks. There was a desire to demonstrate the present state of the inhabited world under Augustan control—not without some criticisms of policy—perhaps for a Greek audience.[62] Romans would hardly need to know about the history and topography of Rome, but might wish to learn more about Anatolia, which, according to Strabo, was the true heart of the civilized world, since it was the location of Troy, the birthplace of Homer, and the home of a steady stream of famous people since that time (more than 200 are catalogued). Greeks, however, would want to know more about Italy and the western Mediterranean, the latter a region that relatively few Greeks might visit. All this territory was now under a common political umbrella but was diverse ethnically and geographically, and a knowledge of the whole would be valuable to all. Strabo stressed the importance of geography to the educated person of power:

> Thus the manifest usefulness [of geography] for political activities and for those of commanders, as well as the understanding of the heavens and things on the earth and sea (animals, plants, and fruits, whatever is to be seen in each place), assumes the same type of man as the one who gives consideration to the art of life and happiness,[63]

a view that betrays the Stoic quality of much of Strabo's education. His emphasis on the importance of geography for the political and military elite was certainly a practical concern in the Augustan world, and perhaps went back to his own family's connection with the erudite Mithridates VI of Pontos, and Strabo's association both with the equally scholarly Juba II of Mauretania and the geographically challenged Aelius Gallus.

Oddly, this most important of geographical and cultural works barely survived. Strabo probably never published it, since it was unknown to assiduous and astute scholars such as Pliny, Plutarch, and Ptolemy. It only seems to have emerged in the second century AD, perhaps with the sole copy making its way to a library in Byzantion, but even then it did not become widely known until Byzantine times. Despite the tenuous nature of its current existence, the importance of the *Geography* to the understanding of the classical world cannot be overestimated, and the present work would have been impossible without it.

CHAPTER 9

THE REMAINDER OF THE FIRST CENTURY AD

The End of Exploration in Northern Europe (Map 9)

Augustus' decision to effect a withdrawal from the lands east of the Rhine after the disaster of AD 9 resulted in little further information about these regions during the rest of antiquity. Germanicus, Drusus' son, was placed in command in Germany shortly before Augustus' death in AD 14. He sailed along the coast of the North Sea (this is probably when that name, Septentrionalis Ocean, came into use) among the Frisian Islands off the modern Dutch and German coasts, identifying 23 of them.[1] The mouth of the Albis (Elbe) had been reached by Tiberius (or a fleet sent by him) in AD 5,[2] and Pliny recorded a long list of toponyms all the way from the Frisian Islands to the Vistula, implying that there was a Roman reconnaissance in the Baltic during these years.[3] The sources are confused, with a mixture of material probably going all the way back to Pytheas, but Augustus himself was explicit:

> my fleet sailed on the Ocean from the mouth of the Rhenus eastward as far as the lands of the Cimbrians, which before that time no Roman had gone into, either by land or sea.[4]

Pliny also referred to this expedition:

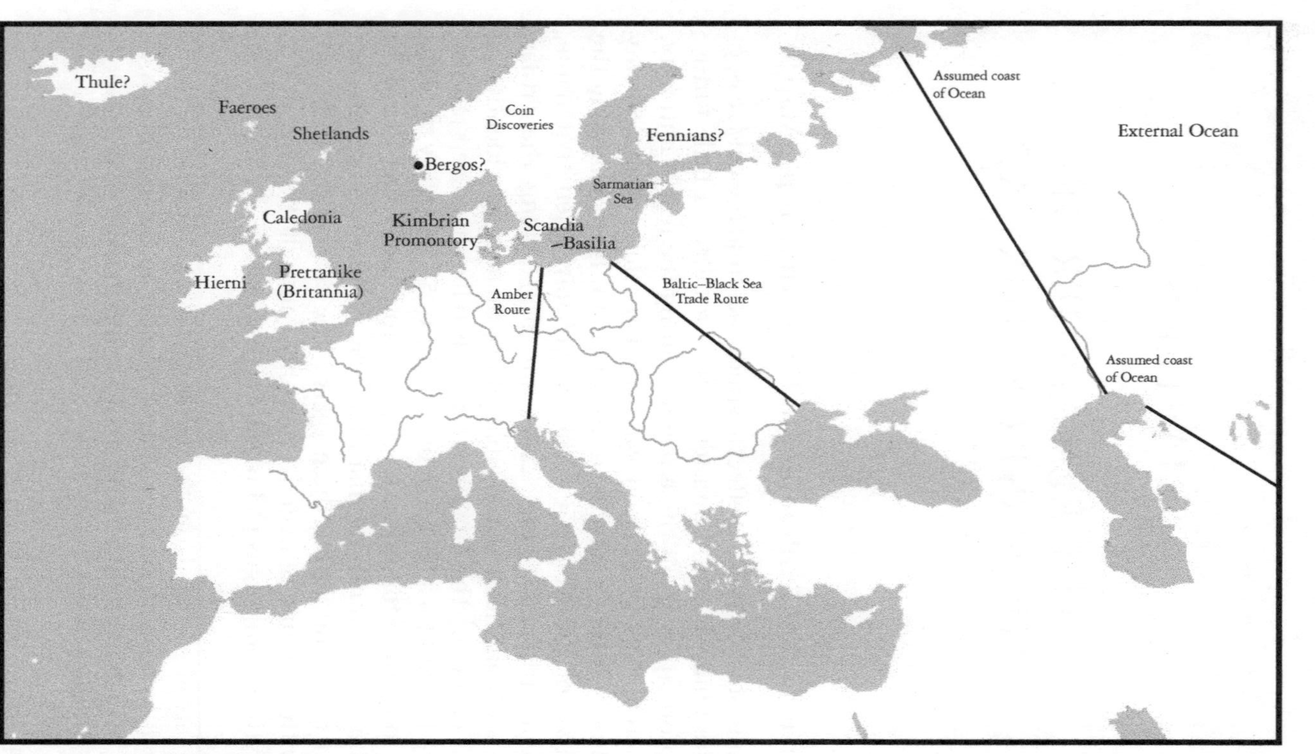

Map 9. Northern Europe.

> The larger part of the Northern Ocean has been sailed on, under the auspices of the Divine Augustus, when a fleet sailed around Germania to the promontory of the Cimbrians, and then—seeing a great sea, or learning about it through report—to the expanse of Skythia, frozen with excessive moisture.[5]

The statement is both explicit and ambiguous. It relates how the Romans rounded the Cimbrian Promontory and came to (or only heard about) a great sea beyond. The Cimbrians (or Kimbrians), perhaps first mentioned by Poseidonios,[6] originally lived on a peninsula in the far north: going past their peninsula and entering another sea suggests that they were on Jutland. The use of the name "Skythian" is probably more generic than specific, the traditional name for the people of the far north, although the comment on the weather seems real.

Enough was known about this region that a certain Philemon wrote a treatise on the Baltic areas, probably in the first half of the first century AD. It may have been a general work on the north, since there was also material on Ireland.[7] His reliance on merchants as well as including information about the amber trade demonstrates what would become the primary reasons for contact between the Mediterranean world and the far north after the Roman military withdrawal. During the reign of Nero the land route north across Europe was refined—for the benefit of those in the amber trade, not the military—by a Roman equestrian who went from Carnuntum (on the Danube below Vienna) to the Baltic, a distance of 600 miles.[8] He probably went up the Morava and reached either the Oder or Vistula, going down to the Baltic.

The Roman government had an additional reason for ceasing exploration of the northern regions, especially by sea. The loss of Varus' legions was serious enough but in AD 16, seven years later, perhaps while he was in the Frisian Islands, Germanicus was afflicted by a storm that destroyed his fleet.[9] Some troops were washed up on inaccessible islands and starved. One of Germanicus' officers, Albinovanus Pedo, wrote a poem about what he had seen, 23 intense lines about the horrors of sea travel:

> The ships are sinking in the mud and the fleet loses the swift wind. They believe that an indifferent destiny leaves them to the wild sea creatures, torn apart through an unhappy fate.[10]

Germanicus was soon recalled by Tiberius, and official Roman activities in the far north of Europe were curtailed. Even Tacitus' *Germania*, written at the end of the first century AD, has little new information and relies heavily on earlier accounts such as Caesar and perhaps Poseidonios, and is centered on regions near the Rhine and Danube. The Albis is hardly known and the Vistula not mentioned at all, although, astonishingly, there is an ethnology of the Fennians (the Finns), from an unknown source and their first appearance in ancient literature.[11]

India and Beyond in Imperial Times

The only part of the Ocean that continued to be traversed regularly in Imperial times was the Indian, as a steady stream of ships went from Mussel Harbor and Berenike on the Red Sea to India, returning with aromatics, spices, gems, precious metals, textiles (including silk), ivory, and pearls.[12] Nothing this valuable went east, and there was an immense imbalance of trade (50 million sesterces annually in Pliny's day).[13] An account of the trade and its routes survives from the middle of the first century AD, known as the *Periplous of the Erythraian Sea*, by an unknown author of Egyptian Greek origin, probably a merchant involved in the trade.[14] This fascinating document is a trove of information, unique in ancient literature, and the primary source on its topic and for the southeastern part of the *oikoumene* in early Imperial times. Two trade routes from the Red Sea are described: one is down the coast of Africa as far as the port of Rhapta (near Dar-es-Salaam), and the other to India, heading to Barygaza on the northwest coast or Muziris farther south. The *Periplous* provides the first detailed information about southern India as far as the straits between the mainland and Taprobane (modern Sri Lanka), where the Red Sea ships ended their run, unable to negotiate the narrows or unwilling to go around the island. But the account also describes the east coast of India—access was by local boats—as far as the mouth of the Ganges. The Roman trading presence was remarkably strong in southern India: at Muziris, on the west coast at or near modern Cranganore, there was a Temple of Augustus, evidence of a local Roman trading establishment,[15] and at Arikamedu, south of modern Pondicherry (perhaps the Poduke of the *Periplous*) excavations have discovered physical evidence of a Roman trading station, perhaps established by the first century BC and lasting until around AD 200.[16]

The author of the *Periplous* also knew about lands east of India (Map 10). Beyond the mouth of the Ganges was Chryse, an island that was the easternmost of all lands, "under the rising sun."[17] The name is a Greek descriptive term—Golden—so it cannot be the actual toponym, but it is very much a specific place to the author. Various suggestions have been made for its location, generally placing it somewhere between the mouth of the Brahmaputra and the island of Sumatra, or perhaps on the Andaman Islands.[18] There was also Thina, obviously China, the first use of the toponym, far to the northeast on the coasts of the Ocean, a location based on an assumption—from as early as the time of Alexander and Patrokles—that the coast ran rather directly from the alleged mouth of the Caspian Sea to somewhere east of India, although the existence of Chryse meant that India could no longer be considered the eastern end of the *oikoumene*. As usual, mention of Thina actually refers to a trade route in that direction, somehow connected with China. Thina is "a great inland city," but there is only the vaguest knowledge of the place itself. There is also a vivid description of an annual trading rendezvous on the border of Thina (here suggesting a territory rather than a city), to which the raw materials for malabathron (an aromatic whose modern equivalent is not

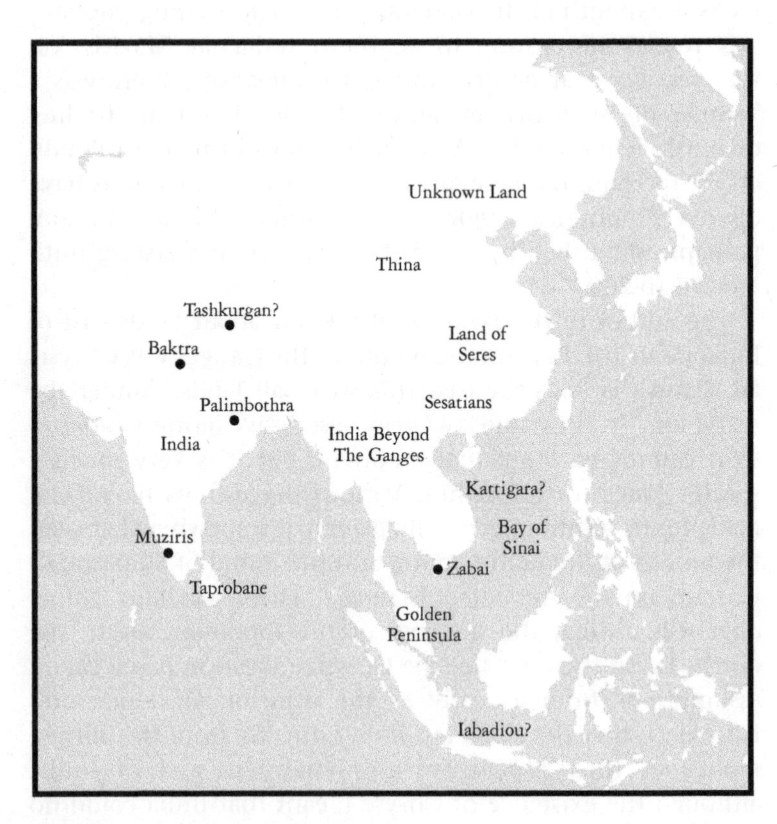

Map 10. Far eastern toponyms.

certainly known, possibly patchouli) were brought by the Sesatians, perhaps a form of the name Sechuan.[19]

After describing the rendezvous, the *Periplous* comes to an abrupt end:

> The places beyond here, because of extremes of storms, great cold, and difficult terrain, as well as the divine power of the gods, have not been discovered,[20]

the only hint in antiquity, faint as it is, of the wilds of Siberia.

Taprobane (modern Sri Lanka), known slightly since the time of Alexander but which the author of the *Periplous* seems to have mentioned only in passing,[21] came to be better understood through the account of a certain Annius Plocamus, who around the middle of the first century AD was a tax collector for the regions of the Red Sea.[22] It seems that he was also involved in the Indian trade, for one of his freedmen was allegedly carried off course to a place called Hippurus, on Taprobane, where he was hospitably welcomed by the local king and stayed for six months. It is probable that Plocamus had commissioned his freedman to investigate new markets. Eventually the king sent envoys to the emperor Claudius (reigned AD 41–54), led by a certain Rachia (perhaps "Rajah"), resulting in a detailed—if somewhat exaggerated—report on the island and its people. In addition, Rachia's father had visited the country of the Silk People, and through his son was able to provide the Romans with further details about them, especially their physical characteristics. They were taller than average, with reddish hair and blue eyes, which does not suggest the Chinese, but Indo-European peoples of central Asia, perhaps middlemen on the Silk Route from India to China, living in the western parts of the Chinese empire.[23] Rachia's father was also aware of the great rendezvous at the borders of Thina.

There was little more learned about India and its region during antiquity, although occasionally Indians appeared in the Mediterranean world, such as an embassy to the emperor Trajan in the early second century AD.[24] India also continued to be an element in popular imagination: Apollonios of Tyana and the apostle Thomas were said to have visited there. Ptolemy recorded some new toponyms in his *Geographical Guide* of the mid-second century AD,[25] but in antiquity there was never again the closeness between India and the west that had existed in Hellenistic and early Roman times.

The Journey to Kattigara

In the era of Alexander the Great, and for some time thereafter, India was the eastern limit of the *oikoumene,* although over the years there was a developing awareness of the land of the Silk People lying somewhere far to the northeast. By the first century AD more data were available. The author of the *Periplous of the Erythraian Sea* knew about Chryse, somewhere far to the east, and Marinos of Tyre early in the following century was able to outline a route based on information from a certain Alexandros (whether a geographer or seaman is not known), which is preserved in Ptolemy's *Geographical Guide.*[26] Chryse, a vague toponym to the author of the *Periplous,* had become a peninsula, confirming that the Malay peninsula was meant. The route of Alexandros (who is otherwise unknown) went from India to the Bay of Sinai, Zabai, and Kattigara, all part of a region generally called "India Beyond the Ganges."[27] After following the coast, the journey made a traverse of a stretch of land, which would suggest a crossing of the Malay peninsula, presumably at the narrows in southern Thailand, and then came to the Bay of Sinai. Marinos and Ptolemy were unaware that the Sinai (or the Chinese) were the same as the Seres (Silk People), now described by a political rather than a commercial ethnym. Thus the Bay of Sinai would be the South China Sea. Eventually Zabai was reached, probably somewhere near the Mekong delta in Vietnam, where Roman goods and coins of the second century AD have been found at the site of Oc-èo, perhaps the best confirmation of the existence of a trade route.[28] From Zabai one continued to Kattigara. The direction is south by east, which would take one to the islands of Indonesia, but the data are confused, and Kattigara may actually be farther north along the coast, perhaps around Hanoi.[29] To the southeast was the island of Iabadiou, whose name implies Java, and which, in one of Ptolemy's rare explanatory comments, meant "barley." It had gold and a town named Argyre (Silver) at its west end.[30] Kattigara and Iabadiou are the last places recorded at the southeast end of the *oikoumene*, and although there was a hint of land beyond, nothing was really known about it.

Romans in and Beyond the Sahara (Map 11)

Mauretania came under the rule of King Ptolemy after the death of his father, Juba II, in AD 23 or 24. A dispute with his cousin, the emperor Gaius Caligula, led to the king being summoned to the imperial presence and eventually executed, sometime during the emperor's short reign (AD 37–41). As might be expected, a revolt broke out in Mauretania, and it took the Romans several years to effect provincialization.[31] As part of this program of stabilization, C. Suetonius Paulinus was sent in AD 41 to investigate the territory. He became the first Roman to cross the Atlas mountains, going beyond them into the tropics, as far as a river called the Ger (probably the Niger, although the name is generic).[32] In one sense this was not new territory, for the trans-Saharan route had been known to Greeks since the fifth century BC, and Juba had made the region more familiar to the Romans, but no Roman had gone that far south. Suetonius Paulinus' report is full of fear and wonder at the flora and fauna, the snow-capped peaks of the Atlas contrasted with regions where it was hot in winter, and volcanic phenomena. The elusive locals contributed to the horror of the region:

> None of the inhabitants is to be seen in the daytime, and everything is quiet because of the terror of the desert, so that an awesome silence comes over those approaching, and, moreover, a terror at what is raised above the clouds into the neighborhood of the lunar sphere. In the nights it flashes with frequent fires and is filled with the wantonness

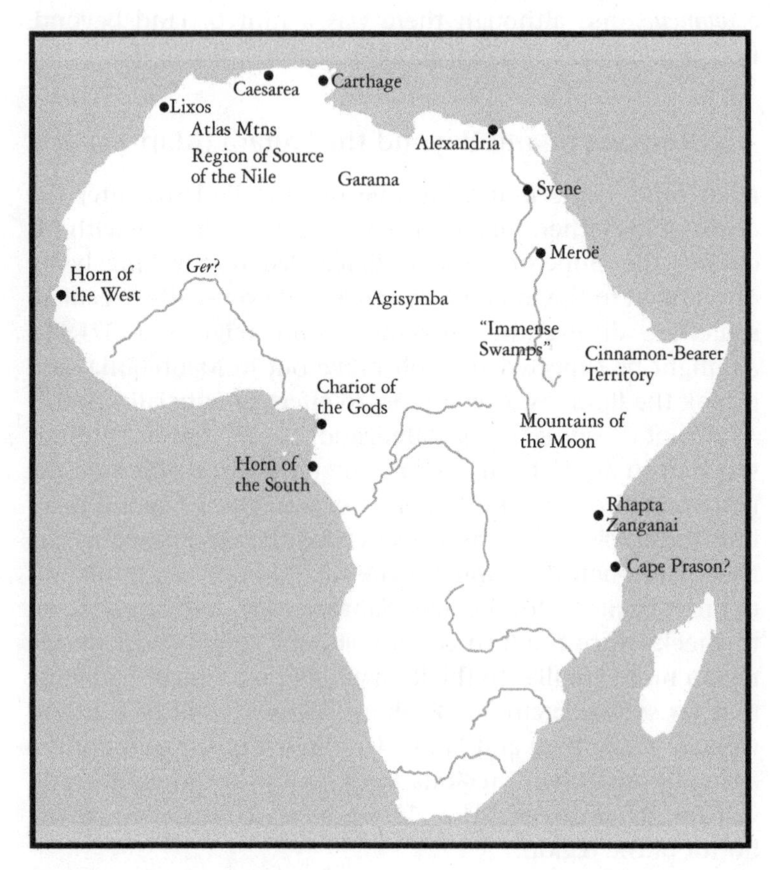

Map 11. Interior Africa.

of Aegipanes and Satyrs, and the music of flutes and pipes as well as the sound of drums and cymbals.[33]

With the provincialization of all north Africa from Egypt to the Atlantic, the Romans became more interested in exploring to the south, as the expedition of Suetonius Paulinus demonstrates. Several journeys were made into and across the Sahara, especially in the latter first century AD.[34] Their focal point was the wealthy oasis district of Garama (modern Germa in the Fezzan district of Libya), about 450 miles inland, noted today for its rich archaeological remains. L. Cornelius Balbus had visited it around 19 BC, and thereafter the Garamantians had a contentious relationship with both Juba and Rome.[35] Later there was an expedition led by Septimius Flaccus, who was not mentioned by Pliny, so he probably traveled after the AD 70s. Flaccus went to Garama and then overland to Aithiopia, taking an additional three months. No other details are provided, and thus it is not certain where he ended up, perhaps at Meroë, but he did record some data that was used by Marinos of Tyre and ended up in Ptolemy's *Geographical Guide*.[36]

Somewhat better known, although also based solely on the sparse data provided by Ptolemy, is the journey of Julius Maternus, probably a few years after Flaccus. He, too, went to Garama, possibly as a Roman envoy or a merchant, and joined the local king on an expedition to Agisymba, which was noted as a rhinoceros habitat. The journey took four months. Agisymba cannot be located: Marinos at first placed it 24,680 stadia south of the equator, but then adjusted this to 12,000, probably based on the length of Maternus' journey, arbitrarily corrected. Even this lesser figure (probably more than 1,000 miles) is absurdly far south—the latitude of Zimbabwe—and the most likely possibility for Agisymba is the region of Lake Chad.[37] No one but Ptolemy mentioned Agisymba, which was considered to be on the parallel of the southernmost limit of the *oikoumene* (thus in theory replacing the parallel of Eratosthenes' Cinnamon-Bearer Territory, but the evidence is unclear). It was in a region of large mountains whose names were not recorded, perhaps the Adamawa plateau that extends southwest to Mt Cameroon, or even Mt Massah east of Lake Chad. Maternus' expedition is the farthest known into central Africa by any Roman: it is possible that he was interested in the rhinoceros trade, since Ptolemy's account of the journey mentioned no other animal, and the rhinoceros had been used for spectacles in Rome since 55 BC.[38]

The Source of the Nile (Map 11)

The upper parts of the Nile continued to excite curiosity and even passion. Julius Caesar had allegedly said that he would give up the civil war if he could only see the source of the river.[39] An academic controversy about the Nile, with accusations of plagiarism, broke out between two scholars in Alexandria, Eudoros and Ariston; Strabo sided with the latter.[40] The reasons for the flooding of the river—debated since the time of Thales—were an ongoing matter of argument, and, it came to be realized, were connected with understanding its source.[41] The Ptolemaic explorers of the early third century BC, with some additional material from Juba, continued to provide the main accounts until the time of the emperor Nero (ruled AD 54–68), who commissioned a major reconnaissance of the upper river, contemplating an attack on southern Aithiopia. His explorers went above Syene, recording distances and toponyms, as well as the change in flora and fauna as one left the desert and entered the tropics. They said that they went 1,996 miles above Syene, an improbable figure that would have taken them well beyond the source of the river.[42] Nevertheless, their journey, supplied from Meroë, "penetrated the farthest points." Nero's tutor and advisor, Seneca, recorded some details, having received an oral report from the leaders:

> After many days we came to immense swamps, whose exit neither the locals knew nor could anyone hope to find, with vegetation so entangled in water. One could not penetrate the water either on foot or by boat, because

> the muddy and overgrown marsh was unable to support anything other than something small with a capacity of one person. We saw two rocks there from which a great force of river water came forth,[43]

perhaps the feature that Seneca elsewhere called the Veins of the Nile, a sacred location in the mountains where water rushed forth.[44] This is the first attempt to connect the source of the river with a specific place on its known course. Needless to say, there have been many futile efforts to locate the Veins of the Nile. The hindering vegetation is the Sudd of southern Sudan, and the Veins must be south of there, perhaps around the Ugandan border, but there is no certain identification.[45] Nevertheless Nero's explorers went farther up the Nile than anyone in ancient times, into regions not seen again by Europeans until the nineteeth century.

The closest that ancient explorers came to the source of the Nile was by a route from the east. Probably around AD 100 a certain Dioskouros recorded that it was a journey of "many days" and 5,000 stadia from Rhapta on the east African coast (around Dar-es-Salaam)—which was the farthest south that had previously been reached—to a place called Cape Prason,[46] probably modern Cape Delgado in northern Mozambique, which Marinos of Tyre believed was on the same parallel as Agisymba and at the southern end of the *oikoumene*. The data are confused, and someone named Theophilos, who actually traveled the route, provided different figures. Despite the contradictions, and indeed the tales of circumnavigation of Africa that had existed for hundreds of years, Cape Prason is the southernmost toponym on the east African coast known from antiquity.

At about the same time, Diogenes sailed along this coast for 25 days, having been driven off course on his return from India, and eventually reached the lakes that were the source of the Nile. This, too, is obviously confused. There is nothing wrong with cruising the east African coast after returning from India because one cannot find the entrance to the Red Sea (this happened to Eudoxos of Kyzikos), but the account implies that the lakes were near the ocean, as if they were coastal lagoons, strangely remindful of Euthymenes' finding the source of the Nile on the coast of west Africa. Elsewhere in the *Geographical Guide*, Ptolemy, more accurately, placed the lakes "quite far inland."[47] A further account, without attribution (but probably also from Diogenes), states that there were mountains near the source of the Nile, whose snow melt fed the river, thus simultaneously solving the twin problems of the origin and the flooding of the river.[48] Ptolemy called these mountains Selene, the fabled Mountains of the Moon of nineteenth-century explorers (the modern Ruwenzori range), whose lakes are indeed the source of the Nile.

These three explorers—Dioskouros, Theophilos, and Diogenes—known only from Ptolemy's synthesis of the geographical work of Marinos of Tyre, present a fascinating yet contradictory picture. They were at the very limits of the world: the name Mountains of the Moon, although probably referring to their rugged landscape, is nonetheless strikingly appropriate. Ptolemy spread them across 10° of latitude (several hundred miles), which is impossible, but one would hardly expect detailed accuracy from unmeasured reports in a remote area briefly visited. The coastal sailing distances are also inconsistent. The lack of confirming evidence about these explorations from any source other than Ptolemy's recension of Marinos would naturally produce unclear data, yet the strongest confirmation of Diogenes' journey is the use of a Greek toponym for the mountains rather than an indigenous name.[49] Nevertheless Dioskouros, Theophilos, and Diogenes went farther south than any specifically documented travelers in antiquity, some of the very few to cross the equator: assuming that Cape Prason is modern Cape Delgado, it lies at 10° 52′ south latitude.

The region of the source of the Nile remained one of the most mysterious of places until modern times, when Europeans began to explore the area, drawn by Ptolemy's Mountains of the Moon.[50] Richard Burton called them the "Lunatic Mountains" due to the obsessive interest they generated.[51] In finding the lakes and mountains again in the mid-nineteenth century, he and John Hanning Speke replicated the journey of Diogenes.

The Romans in the British Isles

Like Mauretania, Britain came under direct Roman control during the reign of Claudius. Caesar's brief visits in the 50s BC, as well as Pytheas' earlier explorations, had made the region known to the Mediterranean world, and trade was extensive enough in the Augustan period that there were four routes to the island: from the mouths of the Rhenos and Sequana (modern Rhine and Seine), and the longer sea voyages from the Liger and Garouna (modern Loire and Garonne).[52] There was an export trade, and Britons were seen in Rome. Its weather was notably unpleasant, with lengthy fogs. Augustus had decided that it was not worthwhile to make Britain a province, as the cost of occupation would be more than its revenue,[53] yet it was claimed that almost the entire island was already under Roman control: an exaggeration, since Roman knowledge in the Augustan period was limited to the southeast.

The Roman attitude changed in the AD 40s. The emperor Gaius Caligula may have planned an invasion, which was suddenly cancelled for reasons that remain obscure. Nevertheless, this indecisiveness led his successor, Claudius, to implement a full-scale takeover, beginning in AD 43.[54] For the next 40 years the military needs of Roman commanders resulted in a remarkably thorough exploration of the island, producing the first important additions to its topography since Pytheas and Caesar. The most complete account comes from the period of the governorship of C. Julius Agricola (AD 78–84), with information recorded in the eulogistic account written by his son-in-law, the historian Tacitus. The work, merely titled *Agricola*, is much more than a biography, and has as one of its themes a sense of remoteness—the typical Greco-Roman view when dealing with the environment of the Atlantic Ocean—but which evolves to the realization that the ocean provides the access to Britain, and thus the Atlantic, like the

Mediterranean, is now part of Rome. Calgaeus, a Caledonian chieftain defeated by Agricola, expressed the sentiments:

> there is no farther land, and the sea itself is no longer secure to us, because of the threat of the Roman fleet.[55]

Needless to say, this is a political statement, not a practical one, and stands in contrast to the hazards of ocean travel faced in the Augustan period.[56]

Agricola went far to the north, well into Scotland, a region familiar to few Romans. He also commissioned a coastal survey:[57] it is not recorded how extensive this was, but it proved that Britain was an island. In addition, the fleet visited the Orkneys.[58] Clearly Agricola was retracing ground covered by Pytheas, and since he had gone to school in Massalia, he may have known more about the explorer than many of his predecessors.[59] Agricola also claimed to have rediscovered Thule.[60] He probably was at Mainland in the Shetlands, and Thule was already moving around the North Atlantic and had become a paradigm for any land in the remote north: even T. Flavius Vespasianus, the future emperor, was said to have determined its location at the time of the original invasion of AD 43.[61] Nevertheless, although it is unlikely that Agricola came anywhere close to Pytheas' Thule, he was the first since him to add significantly to the topographical knowledge of the North Atlantic.

When Agricola was in southwestern Scotland in AD 82, he became aware of Ireland, only 15 miles away. He probably already knew something about the island: it had been on the fringes of Greco-Roman knowledge for centuries. The Carthaginian Himilko may have been the first to hear about it, around 500 BC, and there was a steady stream of references to Ireland in Greek and Roman literature.[62] Whether Pytheas knew about it depends on the interpretation of a statement by Strabo that mentions Ireland (as Ierne) and Pytheas together, without actually associating the two.[63] Yet other details recorded by Strabo, including that it was wretchedly cold and barely habitable, and that those who did live there were among the most savage of people, may be data from Pytheas. Ireland began to emerge from the shadows at the time of Julius Caesar, who reported that it was west of Britain and smaller, and that Mona (the Isle of Man) lay between the two.[64] Pomponius Mela reinforced the negative qualities of the island that had been reported by Strabo (whose *Geography* he did not have access to, suggesting an earlier source) but also added that it was fertile.[65]

With at least some of the available material on Ireland in hand, Agricola considered an invasion. He had taken an Irish chieftain hostage, who may have had his own reasons for Roman support, since Agricola heard far more positive things about the island than had his predecessors. Some confusion about its location—that it was not that far from the Iberian peninsula—put forth by Caesar and perhaps originating from Poseidonios, may have led Agricola to believe that conquest would be useful to Rome. Moreover, there were already Roman merchants on the island. But no invasion took place, although Agricola may have made a brief reconnaissance. Yet half a century later Ptolemy was able to add many Irish toponyms to his account, all coastal, probably reported by the traders and merchants.[66] Nevertheless its exact position and location remained confused until the end of antiquity.

Pomponius Mela

There are two significant Roman geographical authors extant from the mid-first century AD: Pomponius Mela and Pliny. Both are demonstrative of an increased Roman interest in the topic, and Pliny had a certain amount of personal geographical experience, although the career of the former remains elusive.[67]

Mela came from the city of Tingentera in southern Iberia, which was probably on the coast opposite Tingis (modern Tangier). Nothing else is known about his life beyond his work, probably titled *Chorographia*, in three books, and which was written when the emperor was preparing for a British triumph, presumably that of Claudius in AD 43.[68] Thus it was published about 20 years after the completion of Strabo's *Geography*, which, however, was not available to Mela. As the first extant geographical treatise in Latin, the *Chorographia* is both important and problematic. Mela is the earliest to provide a Latin vocabulary for geographical terminology (something that had probably developed in the first century BC), as well as latinized toponyms, which can be difficult to relate to longstanding Greek ones. In terms of theory there is little new in the *Chorographia*, and Mela's view of the *oikoumene* is basically that of Eratosthenes (although not mentioned by name), with the Caspian a part of the Ocean and Meroë the southern limits of the world.[69] Taprobane and Thule are not mentioned. The work follows a unique pattern of moving across the inhabited world, with each of the three books having a particular orientation. After general comments about the continents, Book 1 covers north Africa, Egypt, Arabia, the Levant, Anatolia, and the Pontos, and Book 2 Skythia, Greece proper, Italy, Gaul, Iberia, and the Mediterranean islands. Book 3 is a circuit of the coast of the External Ocean. There are some new toponyms. The work has the character of a *periplous*, especially in Book 3, but without its navigational detail and strict coastal orientation. To some extent Mela was a popularizer, with an interest in myth and little concern for the scientific and mathematical aspects of geography: the suggestion that the *Chorographia* is a piece of juvenilia, even a textbook, has merit.[70] Few of the important figures in the history of geography are cited, and, when they are, like Thales and Anaximandros, they are included merely as intellectual figures within the Greek heritage. Yet it is one of the few extant works to mention either Hanno or Eudoxos of Kyzikos, who were perhaps better remembered in Mela's home region of southern Iberia.[71] Cornelius Nepos, the friend of Cicero, seems to have been an important source for the treatise, but any geographical work of his is lost other than Mela's occasional citations and some by Pliny, who did not have a high opinion of Nepos as a geographer.[72]

Mela was part of a Roman tradition of geography whose antecedents have not been preserved, and which was generally separate from the Greek mainstream. His work is slight, but nevertheless demonstrates the attitudes of the early Roman empire toward geography. It also provides occasional bits of information not available elsewhere, often presented anecdotally. One of the most interesting, also taken from Nepos, is the story of the Indians who had made the northern circuit of the External Ocean and ended up among the Boians, a Gallic people, and who were presented to Q. Caecilius Metellus Celer, proconsul of Cisalpine Gaul in 62 BC.[73] The story is tangled and obviously impossible as it stands. It may merely be a paradigm for the nature of the External Ocean, or the "Indians" may have been from elsewhere, or indeed the tale itself may have been transferred from some other environment (the ethnym Boians is common). Yet it is a good example of the kind of intriguing geographical incident found in the *Chorographia*.

Pliny and his Natural History

A generation after Pomponius Mela, C. Plinius Secundus (AD 23–79), better known as Pliny, wrote the most detailed surviving geographical account in Latin from Roman antiquity, Books 2–6 of his wide-ranging *Natural History*. As is the case with Strabo, the large number of citations in the present work demonstrates the importance of the *Natural History* to the study of geography. Pliny served in both Germany and Iberia, and produced a 20-book history (now lost) of the Roman wars in Germany. He died in the eruption of Vesuvius in AD 79. The 37 books of the *Natural History* cover every conceivable topic: Pliny was more of a compiler than an original scholar.

One of the more interesting parts of the *Natural History* is its list of sources in Book 1. In catalogic form this records all the topics that Pliny discusses, as well as the authors he uses. Those for each book are divided into two lists, Romans and others, a valuable record of little-known writers from antiquity. For example, Book 3, the first topographical one, lists 24 Romans, including familiar names such as Augustus and Pomponius Mela, and obscurities such as Turranius Gracilis, cited only by

Pliny and who seems to have written on Iberia and Africa.[74] This is followed by a list of 13 "externis" authors, including Thucydides and Theophrastos as well as the little-known Kalliphanes, who wrote on Libya. In all, the lists of authors for the five geographical books—more than 250 names, although many are repeated—is the most complete catalogue of geographical writers from antiquity: essentially everyone previously mentioned in the present work, from Homer and Thales to Pomponius Mela, and many known nowhere else except in the pages of the *Natural History*.

The five geographical books of the treatise are a thorough exposition of the world as it was known in the second half of the first century AD. Dates in the AD 70s indicate that Pliny was still working on it in the last years of his life.[75] The section opens with a theoretical outline, relying on sources such as Anaximandros, Pytheas, Eratosthenes, and Poseidonios. Pliny felt it necessary to remind his audience there was compelling evidence that the earth was a sphere, something which might still have been a matter of debate among Romans not thoroughly familiar with Greek scientific writings.[76] He also lamented the decline of original research in his era.[77] Pliny's geography is very much the geography of the Roman world, even more so than that of Pomponius Mela.[78] It was designed for the highest levels of the Roman elite, a geographical validation of the importance of the Roman Empire.[79] The dedication is to Pliny's long-time friend Titus, the son of the emperor Vespasian, who would become emperor himself two months before Pliny's death.[80] Much of the treatise was designed to support and affirm the validity of the Roman Empire of his day, and many of the rubrics in the work are fitted into this thesis.[81] Geography, needless to say, was one topic that meshed well with this concept.

Topography was a necessary tool in understanding Rome's place on the earth, and there is much of a traditional *periplous* in Pliny's account, but it goes beyond a mere coastal survey, emphasizing rivers and mountains as an essential part of Roman topography. The rivers were the modes of access to the interior lands under Roman control: a long catalogue of

German rivers is a case in point, and Gaul was also defined by its rivers. The source of the Nile was of particular interest, with Pliny favoring Juba's Mauretanian theory.[82]

The geographical account begins at the Pillars of Herakles, and moves through the known world, basically west to east, yet like Pomponius Mela emphasizing the more central regions before discussing the perimeter. Often there is more detail than in earlier sources: the report on Taprobane is the fullest to date.[83] Elsewhere there are immense catalogues of toponyms and ethnyms, often without elaboration. In Italy and the west these lists may derive from official registers from the Augustan period,[84] but in the eastern provinces the sources go back to Hellenistic times, such as the 70 places and peoples (in two versions, from Bion and Juba) on the Nile around Meroë.[85] As with Pomponius Mela, these latinized recensions of toponyms originally published in Greek (and often taken from a third language), for which the nominative may not be preserved, can be baffling to the modern reader.

The geographical section of the *Natural History* is difficult to understand, in part because it is a Latin rendition of a discipline historically written in Greek. Nevertheless it remains one of the most important topographical surveys surviving from antiquity, equal in value to Strabo and Ptolemy. One may be overwhelmed by the endless lists of toponyms, so many of which cannot be specifically located, but there is also much ethnography and descriptive topography. A useful feature is an interest in comparative distances: figures from several different sources are regularly compared, from Eratosthenes to the map of Agrippa. These are converted to miles, which can raise its own problems, and the numerals in the text can be questionable, but the data are a valuable repertory of the known distances across the *oikoumene*, something of great importance to the Roman self-image.

Plutarch

Plutarch of Chaironeia (*c.*AD 50–120) was not a geographer, but he preserved many geographical details in his works, especially his biographies of people such as Alexander, Pompey, and Caesar, and thus is a major source for their topographical efforts. His essay, *Concerning the Face that Appears on the Globe of the Moon,* includes the speculation that there might be other continents.[86] Perhaps satirically, he also compared the earth to a dining table, pointing out a parallel between it and the abundance of the world.[87]

CHAPTER 10

THE LATER ROMAN EMPIRE

Rome and China (Map 12)

In AD 97, the commander of the western region of the Chinese empire, Ban Chao, sent an envoy, one Gan (or Kan) Ying, to the west.[1] As early as the second century BC, the Chinese had become aware of Greek remnants in Baktria,[2] and in the following century the Chinese had been in contact with the Parthians on the Iranian plateau, learning about the lands farther west, including the notable state of Daqin, the westernmost in the world, which could take several months to reach.[3] Many toponyms along the route are mentioned in Gan Ying's report, which are almost impossible to equate with known ones, except perhaps the capital of Daqin, Andu, which is remindful of Antioch (on-the-Orontes), where one might expect a traveler from the far east to reach the Mediterranean. The Chinese would have good reason to believe it was a great city: it had been the Seleukid capital for many years before becoming the capital of the Roman province of Syria, and it was where Indian envoys met Augustus in 20 BC.[4]

Gan Ying never made it to Daqin, although he knew something about the regions in the west beyond what he actually covered. The Parthians may have discouraged him from going by land, and the sea route was exceedingly long.[5] Nevertheless this is the first documented example of someone from China attempting to reach the Mediterranean. The

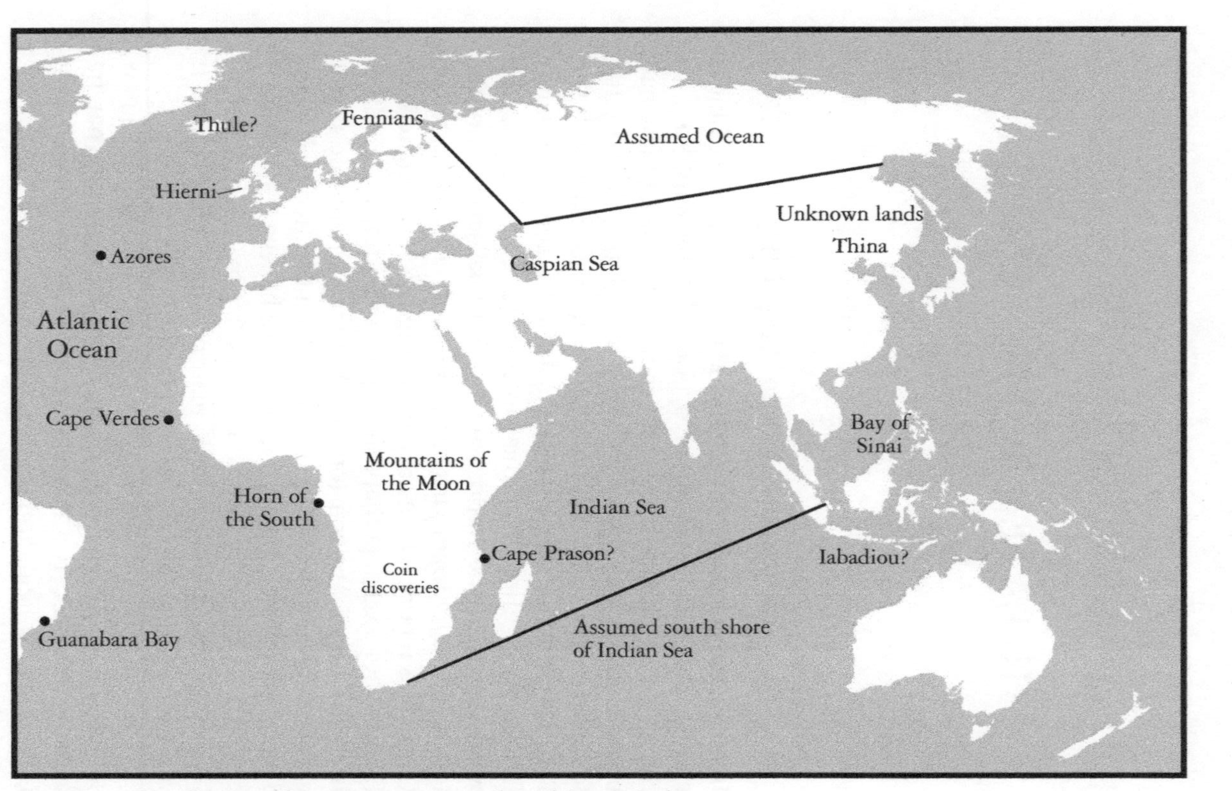

Map 12. The limits of ancient geographical knowledge.

Chinese and the Mediterranean world had had faint knowledge of each other since perhaps the third century BC, but there had been no known previous attempt to make direct contact.[6] Nevertheless, the epitome of Chinese material goods, silk, may have come to the Mediterranean as early as the first century BC and had become relatively common by the following century: as "Seric cloth" it was mentioned more than once by the author of the *Periplous of the Erythraian Sea*.[7] Even as early as the time of Tiberius, silk had become enough of a symbol of ostentatious luxury—at least to that frugal emperor—that legislation was passed forbidding men to wear it, a stricture that the emperor Gaius Caligula violated.[8]

At about the same time as Gan Ying's journey, a Macedonian named Maes Titianus attempted to determine the distance to China from the known world.[9] As usual, details are sparse, but Maes' agents (he did not make the journey himself) set forth from a place called the Stone Tower (Lithinos Pyrgos), 26,280 stadia east of the Euphrates, north of the Imaos mountains and well east of Baktra.[10] The name Stone Tower is common in this region, and even though none of the places so called can be equated exactly with the point of departure of Maes' expedition, the town of Tashkurgan (also meaning Stone Tower) on the western border of China is an intriguing possibility.[11] Stone Tower was obviously a flourishing trading emporium, where merchants from all directions met. Maes' agents eventually ended up among the Silk People, at a location another 24,000 stadia to the east and lying on the parallel of Rhodes. The distance from Stone Tower to the Silk People, essentially the same as from the Euphrates to Stone Tower, is presumably a calculation based on the reported seven months for the journey. Where the Silk People congregated cannot be determined—no place is named—but it may be at the contemporary Chinese capital of Luoyang.[12] The placement of the Silk People on the parallel of Rhodes probably means a member of Maes' expedition determined the latitude of their city, and the parallels of Rhodes and Luoyang are a mere 1.5° apart.

Thus merchants, not diplomats, sustained the contact between the Mediterranean and China. According to Chinese reports envoys from King An-tun reached China and the emperor Huan-ti in AD 166.[13] An-tun is an Antonine emperor, presumably Marcus Aurelius or Lucius Verus, but the failure of the embassy to be mentioned in Greco-Roman sources suggests that it was less official than its partipants implied—probably merchants who pretended to be official functionaries. Nevertheless the Chinese recorded it as the first direct contact between Rome and China, although they were not particularly impressed with the gifts that they received—ivory, rhinoceros horn, and tortoise shells—which they considered to be tribute.[14] Despite the fact that no official relations were ever established between the two states, the merchants on the Silk Road continued to ply their trade, and brought knowledge of each civilization to the other. It is interesting that the alleged embassy from An-tun to the Chinese is at the very time of the first thorough and well-informed description of silk and silk worms in Greek literature.[15]

Marinos and Ptolemy

The final flourishing of the ancient geographical tradition is the *Geographical Guide* of Ptolemy of Alexandria, written toward the middle of the second century AD. It is an unusual and highly influential work, not truly a geographical treatise but one on cartography. There are certain parallels with the *Against the "Geography" of Eratosthenes* of Hipparchos from 300 years previously, since the *Geographical Guide* was also written by a mathematician and astronomer whose primary goal was to use his own disciplines to correct geographical data and the work of a predecessor. In Ptolemy's case this was Marinos of Tyre, one of the most elusive figures in the history of Greek geography, known only from the *Geographical Guide*. He wrote around AD 110, shortly after the Dacian campaign of the emperor Trajan.[16] About a generation later, Ptolemy took Marinos' work and refined it, with the intent of providing instruction on how to create a map of the world.[17] Ptolemy had already published his *Mathematical Syntaxis* (popularly known today by the Arabic form of its name, *Almagest*), which set forth the motions of the heavenly bodies and became the standard work on its topic until the Renaissance. It had a geographical component, since astronomical observations in various places had to be reconciled with one another, and this may have led Ptolemy to a broader work on the location of places in the *oikoumene* and to create (in theory) a map.[18]

The *Geographical Guide* located more places than any previous author. Ptolemy used Eratosthenes as his basic framework, but created a more extensive *oikoumene* than had been known, and was perfectly willing to acknowledge that many of his data were taken directly from Marinos.[19] Marinos had access to a variety of information—Ptolemy praised him as a diligent researcher—and his work had already gone through several editions. Marinos seems to have been particularly sensitive to recent data provided by merchants, but in fact few sources of any type are named in the *Geographical Guide*: in addition to Marinos there are only 11 others, most of which are known nowhere else and are from the generation previous to Marinos (as they were not mentioned by Pliny). Thus they provided new information about the extremities of the inhabited world. Ptolemy and Marinos were not writing geographical works but cartographic manuals, and to them the history of toponyms was less important than determining their location. It is perhaps no accident that of all the luminaries in the history of geography, no one other than Hipparchos is cited in the *Geographical Guide*, the other mathematician and astronomer to venture into geography, whom Ptolemy rightly described as the only one to provide astronomical data for localities.[20] Like Hipparchos, Ptolemy realized that astronomy was the only way to determine an accurate position for places, yet he also fell into the same trap as his predecessor: there simply were not that many locations whose positions had been determined astronomically, and many figures were obtained by a conversion of stadia distances along a route into a longitude and latitude, probably using existing maps.[21]

Ptolemy wrote that 26 maps of the *oikoumene* were created. Whether this was actually done, or the *Geographical Guide* was merely an instruction manual, has long been debated.[22] The maps in the extant manuscripts show little consistency and are no earlier than AD 1300, and some manuscripts have none. It seems improbable that Ptolemy actually included maps in his treatise, and in all likelihood the extant mapping tradition was developed by the Byzantine scholar, Maximos Planudes, in the late twelfth century, who claimed to have rediscovered the *Geographical Guide.* But it was certainly possible for any competent reader to create maps based on the data that Ptolemy provided, something that may have happened as early as the later Roman period.

The *Geographical Guide* consists of 8 books, and thus is second only to Strabo's *Geography* in length. The first and last are theoretical and technical material, with a discussion of topics such as the discipline of cartography, the greater validity of astronomical data rather than travel records, a critique of Marinos, and the actual instructions on how to make a map. Some of the material is descriptive, especially the information from recent explorers that extended the limits of the *oikoumene* in Africa and east Asia. There is also a list of parallels: Eratosthenes' eight have become 23, all defined by length of daylight and degrees of latitude, with practically no toponymic material. The middle six books of the *Geographical Guide* are the detailed lists, with over 8,000 toponyms (Eratosthenes had about 400) each accompanied by a latitude and longitude. Descriptively, the catalogue is sparse (the handful of sources named are all confined to Book 1), and the modern reader can be overwhelmed by the sheer quantity of names. Many of them are new. There are groupings of regional toponyms, with each section introduced by a catalogue of local ethnyms, and an occasional extraneous comment on ethnology, but these are rare.

The addition of data from the far east meant that the *oikoumene* had grown since the time of Eratosthenes,[23] and although Ptolemy reduced some of Marinos' figures in both the east–west and north–south directions, he nevertheless ended up with an inhabited world that extended from the Fortunate Islands (probably the Canaries) in the west, which marked his prime meridian, to the Silk Peoples, 180° to the east or halfway around the globe, nearly twice the 70,000 stadia that was originally believed to be the east–west extent of the *oikoumene*. This had a major effect on Renaissance explorers, for these figures (and others supplied by Marco Polo) allowed Columbus to believe that the journey from the west side of the inhabited world around to the east was only about one-quarter of what it was,[24] and by 1498, at the time of his third voyage, he thought that he had reached Ptolemy's Golden (modern Malay) Peninsula.[25]

There are many new toponyms in the *Geographical Guide*, especially at the extremities of the world. Their density can be astonishing, something that the modern reader can easily see by looking at the maps created from the text. There are dozens in Arabia and nearly 50 on Taprobane, regions hardly known a century before Ptolemy.[26] Yet in many other areas there are few that are new: parts of northwestern Europe hardly go beyond what was known in the Augustan period. Moreover, there are serious problems with the orientation of Scotland, which is pulled far to the east. Ever since the rejection of Pytheas' data—a process that began in the second century BC—there had been difficulties in considering the far northwest of the world, and Ptolemy's issues represent their culmination.[27] This is a contrast to the large amount of new data in the south and east of the *oikoumene*, reflecting the greater importance of trade with central Africa and the far east than with the Baltic. Ptolemy was the first in many years to enclose the Caspian Sea—although its status continued to be debated until medieval times—and there are some new rivers that flow into its north end, the Rha, Rhymnos, and Daix (the Rha is the Volga and the Daix is perhaps the Ural), which left little room for a connection to the Ocean.[28] On the other hand, Ptolemy continued to believe that the Iaxartes flowed into the Caspian and he had no knowledge of the Aral Sea, so he remained to some extent still dependent on the conception of this region dating back to the time of Alexander and Patrokles.

Yet the most peculiar aspect of Ptolemy's topography was his idea that the Indian Ocean was an enclosed sea. This was a belief that perhaps originated with Hipparchos and was considered plausible by Polybios.[29] Hipparchos used Seleukos of Seleukeia and his work on the tides as his authority, but it seems to have been more of a theoretical suggestion than any actual belief—and contradicted the many tales of the circumnavigation of Africa—until Ptolemy placed it in the mainstream of geographical thought. He even suggested that the Aithiopians were not far from Sinai and Kattigara (east of India). With the authority of Ptolemy, the idea of an enclosed Indian Ocean remained a hindrance to exploration until the time of Vasco da Gama.[30]

Beyond the *Oikoumene*

Ptolemy established the complete extent of the inhabited world as it was known in antiquity. In fact, he made it too large, because of the repeated extensions due to new data received from the trade network of the Roman empire. He was not a geographical theorist and showed no interest in whether there was anything beyond what was already known, and nevertheless any thought of what lay outside the *oikoumene* was the concern of allegorists, not geographers, either practical or theoretical. Yet even Ptolemy's expanded inhabited world was still realized to be only a small part of the sphere of the earth. Cicero, who dabbled in geography and even planned to write a treatise on the topic,[31] had perhaps the best statement on the matter, with Scipio Africanus in the famous dream sequence of the *Republic* describing the limits of Roman power:

> the entire land that you hold is narrow vertically and wider in width, but is a small island surrounded by that sea which you on earth call the Atlantic, the Great, or the Ocean. You can now see how small it is, despite having such a name. Do you believe that your fame, or any of us, could go beyond the settled and known lands, by crossing the Caucasus or swimming across the Ganges? Who among those of the rising or setting sun, or the extreme north or south, will ever hear your name? Cut all these

> out, and you will not fail to see over what a narrow region you are so eager to spread your glory.[32]

To be sure, this is not geography, yet it shows an awareness of the minuteness of the *oikoumene* compared to the entire sphere of the earth, and thus the insignificance of the known world. In fact, there could be people on the opposite side of the earth, who would be upside down relative to those in the known world, and great expeditions would be needed to reach those lands. There might even be people on the moon, an idea perhaps originating as early as Xenophanes of Kolophon. This raised the question of the position of human beings both on the earth and in the entire cosmos.[33] But this was all speculation, even if some of it were later shown to be true, and there is no evidence that the Greeks or Romans had any specific knowledge of a world beyond the limits defined by Ptolemy.

The end of the world in the north was, of course, Thule, which Ptolemy called an island "at the end of the known sea," implying that there might be something more beyond (Map 12).[34] He knew the name Skandiai, an island in the western Baltic, but had no awareness of the Scandinavian peninsula.[35] Ptolemy called the Baltic the Sarmatian Ocean, and beyond it was "unknown land."[36] This shows that the exact position of the Ocean was not really understood north and east of the Baltic—in fact all the way to Kattigara—and that there was more land to the northeast, a contrast to the belief of Alexander and the Hellenistic explorers of the region beyond the Caspian, who always thought that the Ocean was just out of sight. The author of the *Periplous of the Erythraian Sea* had hinted at more land to the north and northeast and knew that it was exceedingly wild.[37] Yet in antiquity there was no sense of what was beyond a line roughly from modern St. Petersburg to Beijing.

Nevertheless trade goods could go beyond the *oikoumene* (Map 12). A few items penetrated into southern Norway and Sweden in the first and second centuries AD, presumably via routes from the islands of Bornholm and Gotland (probably

Ptolemy's "Skandiai") up the east and west coasts of Scandinavia and into southwestern Finland.[38] Roman coins have also been found in Iceland as well as in extreme southern Africa and Madagascar.[39] Trade goods were discovered on the Mekong River in Vietnam at Oc-éo and P'ong Tuk,[40] and there is the curious incident of the Roman amphoras near Río de Janeiro.[41] Yet portable objects travel far, and there is no evidence that these demonstrate any direct contact between the Romans and the far reaches of the world.

The End of Ancient Geography

After Ptolemy the creative era of ancient geography was virtually over, although earlier works were still being analyzed. An odd occurence was the emergence of the *Geography* of Strabo, which, it seems, had never been published after its author's death around AD 24, and which was unknown to Pliny, and probably to Marinos and Ptolemy, although the source citations in the *Geographical Guide* are so few that this is difficult to prove. Yet by the end of the century Strabo's treatise was known to scholars, since Athenaios quoted it twice, and from that time it was regularly cited.[42]

The self-image of the Roman world meant that in later times there was an emphasis on map-making, the most obvious way of displaying the extent of Roman power, much more effective than any literary account.[43] Presumably the map of Agrippa, created at the end of the first century BC, continued to be on display in the Porticus Vispania in Rome, but the total lack of any evidence beyond the literary, as well as no knowledge whatsoever of the porticus itself[44] means that any detailed visual sense is almost impossible to obtain. In the first decade of the third century AD a public map of the city of Rome was created by the emperor Septimius Severus and attached to a wall in the Temple of Peace, which had recently been restored.[45] Fragments of this marble plan (the Forma Urbis Romae) have been recovered since the sixteenth century, and it is the subject of an on-going restoration. It was exceedingly large (60 by 42 feet), and, even though merely a town plan rather than a view of the world, it is the best preserved example of a map actually from antiquity.

The most famous extant world map from the Roman period is the Peutinger Map, which exists today in a parchment copy of the early thirteenth century (or slightly earlier) in the National Library in Vienna.[46] It was a long strip (now 11 separate sheets), 13 inches high and 22 feet long, the ultimate expression of an *oikoumene* longer than it was wide. It is based on an original from the pre-Christian era, and the recent suggestion that it might have been a display piece in Diocletian's palace in Spalatum (modern Spalato or Split in Croatia), and thus from the end of the third century AD, is compelling.[47] Even though the extant copy is a thousand years removed from that era, it nevertheless is a striking visual record of how the Romans conceived their world. The left (western) end is lost (the extant map begins with Gaul and the southeastern British coast), but from this point to India and Taprobane the surviving 11 sections lay out the world of Rome for all to see. Rome itself is exactly at the center (assuming three lost sections in the west), as one might expect. The unusual shape of the map was designed to show the east–west magnitude of the Roman world, presenting an *oikoumene* more than 20 times as long as it is high, and containing virtually no information about the extremities of north or south. It is also somewhat beholden to a longstanding tradition of depicting routes in straight lines, a geographical rule as early as Eratosthenes' meridians and parallels and still existing today.[48]

The Peutinger Map is geography for political leaders, as Strabo had promoted:[49] only the important parts of the inhabited world are shown, a smaller *oikoumene* than that of Ptolemy, limited to those portions that would be important to people in power. Yet it remains the best way to visualize the world as the Roman elite saw it.

Christian Topography

The end of the ancient geographical tradition is marked by the Christianization of the Roman world, and the imposition of a new set of values. The Jewish antecedents of Christianity had long since made their way into Greek thought: Herodotos was perhaps the first Greek to write about the Jews.[50] In Strabo's *Geography* there is a lengthy discussion of Moses—probably derived from Poseidonios—who is depicted as an Egyptian priest who came to Judaea with his followers.[51] However tangential this may be to Strabo's main theme—probably a reflection of his deep interest in cults—nevertheless the passage provides a connection between Jewish culture and Greco-Roman geographical scholarship, something that passed to the early Christians.

Some hints of the change toward a theological view of geography were put forth by Augustine in the early fifth century AD. He was not totally certain that the earth was a sphere—although willing to admit to the possibility—and rejected on biblical grounds that there could be any inhabitants on the opposite side of the Ocean, or any communication with such places. Moreover, he believed that Greek intellectuals such as Plato, no stranger to geographical theory, had been influenced by biblical thought.[52] Ideas such as these marked a major shift in how the world was understood, and thus would come to affect geography.

In the sixth century AD, a merchant from Alexandria named Kosmas traveled throughout Aithiopia and perhaps as far east as Taprobane.[53] From these journeys he gained the surname Indikopleustes (Indian Sailor). He then seems to have become a monk and thereafter devoted his life to writing, composing a geography (now lost) and a surviving work titled *Christian Topography*.[54] His voyages had taken him on both the Mediterranean Sea and the Indian Ocean, and down the coast of Africa. He had also visited Axum (in modern northern Ethiopia),[55] which had emerged in the first century AD and eventually replaced Meroë as the most prominent city of the region.[56] His visit to Axum was in the fifth year of Justinian (AD 522), and he was writing the *Christian Topography* 25 years later, or around AD 547.

Despite its interesting comments on Aithiopia and the far east in late antiquity, the *Christian Topography* is primarily a work on theology, not geography. It sought to use biblical evidence both to deny that the earth was a sphere and to understand the cosmos. There is an elaborately constructed Christian universe, based on an understanding of the sacred furniture that Moses made in the wilderness.[57] Since the Ocean could not be navigated, there was little point in seeking what was beyond it. An inventive concept of the path of the sun fit into this new cosmos, including the suggestion that it disappeared behind a mountain in the far north, and that it was no larger than the distance from Alexandria to Rhodes.[58]

The *Christian Topography* is not to be categorically dismissed, despite a world view that is difficult to comprehend, to say the least. Some of the details of Kosmas' cosmology were from the ancient Greek tradition: the sun and its mountain in the far north are remindful of Anaximenes and Pytheas, and the rejection of sailing the Ocean recalls the poem of Albinovanus Pedo. Other details are new, such as some of the material on Pytheas.[59] Kosmas considered some topographical matters peculiar to the Judeo-Christian tradition, including the question of where Moses and the Israelites crossed the Red Sea (at Klysma, formerly Kleopatris, near modern Suez):[60] there may have been more about such issues in his lost geography. But on the whole the work represents a new outlook, a rejection of a thousand years of Greco-Roman geographical thought. The primacy of classical geographical theory was not to be restored until the publication of the major texts in the Renaissance: Strabo and Pliny in 1469 and Ptolemy in 1475.[61] Although these early editions were generally textually unreliable, and Ptolemy and Strabo were first printed in Latin, they were available to early Renaissance explorers and humanists, and restored ancient views of the world, as well as inspiring the future that would be represented by Columbus and Copernicus.

APPENDIX 1

THE MAJOR SOURCES FOR GREEK AND ROMAN GEOGRAPHY

Geographical comments pervade Greek and Latin literature, and the following list is not meant to cite all the personalities who mentioned geographical issues, but is merely a catalogue of the major explorers and writers. Many of them are known only derivatively, especially from the works of Strabo and Pliny. No distinction has been made between professional geographers, such as Strabo, and writers in other fields whose treatises include a large amount of geographical data, such as Aeschylus or Herodotos. In addition, many explorers did not produce published reports. For a complete list of all known ancient geographers, see *EANS* 999–1002.

Aeschylus (early fifth century BC), the famous Athenian tragedian, included extensive geographical data in his plays.
Agatharchides of Knidos (early second century BC) wrote an historical account of the Red Sea region that included a large amount of geographical material.
Agathemeros, of uncertain Roman date, was the author of a brief summary of the history of geography.
Agricola, C. Julius (AD 40–93) explored Scotland and sent a fleet around most of Britain, in the early AD 80s.
Agrippa, M. Vipsanius (*c.*62–12 BC) created the first large-scale public map of the Roman world.

Anaxikrates (of Rhodes?) (latter fourth century BC) made an exploration of the Red Sea.
Anaximandros of Miletos (early sixth century BC) theorized about the shape of the earth.
Anaximenes of Miletos (mid-sixth century BC) considered the effect of natural phenomena on the earth.
Androsthenes of Thasos (late fourth century BC) explored the Persian Gulf.
Antiochos of Syracuse (fifth century BC) wrote about the history of Italy (to him, only the far south of the Italian peninsula) and Sicily.
Apollonios of Rhodes (latter third century BC) included much geographical data in his *Argonautika*.
Archias of Pella (late fourth century BC) explored the Persian Gulf.
Aristeas of Prokonnesos (early seventh century BC) traveled north of the Black Sea, and may have written a poetic account of his journey.
Aristokreon (early third century BC) wrote an *Aithiopika*.
Ariston (third century BC) cruised the Red Sea.
Aristotle (384–322 BC) wrote about geographical issues in his *Meteorologika* and other works.
Arrian (L. Flavius Arrianus) (*c.*86–160 BC) wrote *Anabasis of Alexander* and *Indika*, containing the fullest extant information about both those topics.
Artemidoros of Ephesos (*c.*100 BC) wrote a *Geography* in 11 books.
Bion of Soloi (early third century BC) traveled up the Nile and wrote an *Aithiopika*.
Daimachos of Plataiai (early third century BC) was Seleukid ambassador to the Mauryan court, whose *Indika* was not much respected.
Dalion (early third century BC) sailed up the Nile beyond Meroë and wrote an *Aithiopika*.
Demodamas of Miletos (early third century BC) went beyond the Iaxartes River on the commission of Seleukos I, and wrote an account of that region.

Demokritos of Abdera (fifth century BC) wrote a *Kosmographia,* and believed that the inhabited world was oblong in shape.
Dikaiarchos of Messana (late fourth century BC) devised a method for measuring the heights of mountains, and created the prime parallel of the inhabited world.
Diogenes (*c.*AD 100) may have discovered the source of the Nile and the Mountains of the Moon.
Dioskouros (*c.*AD 100) went down the east African coast to Cape Prason.
Drusus, Nero Claudius (38–9 BC) explored east of the Rhine and north of the Alps.
Ephoros of Kyme (mid-fourth century BC) wrote the first universal history, which included a section on geography that divided the inhabited world into ethnically based quadrants.
Eratosthenes of Kyrene (latter third century BC) invented the discipline of geography, including the term itself.
Eudoxos of Knidos (mid-fourth century BC) refined the theory of zones.
Eudoxos of Kyzikos (late second century BC) established the route to India, and attempted to reach it by circumnavigating Africa.
Euthymenes of Massalia (*c.*500 BC) traveled down the coast of west Africa and developed a theory about the course of the Nile.
Hanno of Carthage (*c.*500 BC) traveled down the coast of west Africa at least as far as Mt Cameroon. A summary of his report in Greek is the earliest surviving *periplous.*
Hekataios of Miletos (*c.*500 BC) wrote a *Circuit of the Earth,* the first such work in Greek literature.
Herodotos of Halikarnassos (*c.*485–420 BC) wrote the first extensive prose history, which includes much geographical information.
Hesiod (*c.*700 BC) was the first to show awareness of the Black Sea and the major rivers of Europe.
Hieron of Soloi (late fourth century BC) explored the Persian Gulf and the Arabian coast.

Himilko of Carthage (*c*.500 BC) explored the Atlantic north of the Pillars of Herakles, perhaps reaching Ireland.
Hipparchos of Nikaia (second century BC), a mathematician, was the first to blend astronomy and geography.
Homer (late eighth century BC), the epic poet, included the earliest extant Greek information on places and peoples in his poems, and was widely revered by later geographers.
Juba II of Mauretania (*c*.46 BC – AD 24) explored northwest Africa, discovered the Canary Islands, and wrote about the southern extent of the known world.
Julius Caesar (100–44 BC) explored Gaul, and was the first Roman to visit Britain and cross the Rhine.
Julius Maternus (*c*.AD 80s) went into north central Africa, perhaps to Lake Chad.
Kolaios of Samos (late seventh century BC) made the first known Greek voyage outside the Pillars of Herakles, reaching Tartessos in southwestern Iberia.
Kosmas Indikopleustes (first half of the sixth century AD) wrote the first Christianized geography.
Ktesias of Knidos (*c*.400 BC), wrote a history of Persia, of uncertain value, but with much geographical data about the eastern part of the known world.
Marinos of Tyre (early second century AD), a mathematical cartographer, was the primary source for Ptolemy's *Geographical Guide.*
Megasthenes (*c*.300 BC), Greek ambassador to the Mauryan court, wrote an *Indika*, which became the standard work on the topic.
Metrodoros of Skepsis (*c*.100 BC), the court historian of Mithridates VI of Pontos, wrote on the Baltic.
Midakritos (of Massalia?) (perhaps sixth century BC), went beyond the Pillars of Herakles in search of tin.
Nearchos of Crete (late fourth century BC) was commander of Alexander's fleet on the Indus and during the return to the Persian Gulf, and wrote a report of the cruise.
Onesikritos of Astypalaia (late fourth century BC) was the pilot of Alexander's fleet on the Indus and during the return to the Persian Gulf, and wrote a report of the cruise.

Parmenides of Elea (early fifth century BC) developed the theory of terrestrial zones.

Patrokles (early third century BC) explored the Caspian Sea for Seleukos I.

Periplous of the Erythraian Sea (mid-first century AD) whose author is unknown, describes the routes to India and along the coast of east Africa.

Philemon (probably first half of first century AD) wrote on the Baltic.

Philon (early third century BC) explored both the Red Sea and the upper Nile for the early Ptolemies, perhaps seeking a route between the two.

Pindar (early fifth century BC), the Boiotian poet, included a surprising amount of geographical data in his poems.

Plato (427–347 BC), the Athenian philosopher, considered the size of the earth, and with his description of Atlantis originated fantasy geography.

Pliny the Elder (AD 23–79) is one of the primary sources for ancient geographical data, in his lengthy *Natural History*.

Polybios of Megalopolis (*c.*180–118 BC) explored the west African coast and perhaps north to the mouth of the Loire, and theorized about the nature of the equatorial regions. He included much geographical data in his *Histories*.

Polykleitos of Larisa (latter fourth century BC) seems to have been the recorder of Alexander's topographical manipulations.

Pomponius Mela wrote the first extant geography in Latin, in the AD 40s.

Poseidonios of Apameia (*c.*135–50 BC) wrote *On the Ocean*, an account of the far western areas of the inhabited world, and the first detailed Keltic ethnography.

Pseudo-Skylax is the name given to the author of a *periplous* of the Mediterranean, Black Sea, and part of the adjacent Atlantic of the 330s BC.

Ptolemy of Alexandria (mid-second century AD), a mathematician, wrote the *Geographical Guide*, a cartographic guide that became the standard work on the topic through medieval times.

Pythagoras (early third century BC) explored the Red Sea for Ptolemy II.
Pythagoras of Samos (sixth century BC), the philosopher, may have been the first to suggest a spherical earth.
Pytheas of Massalia (latter fourth century BC) explored the north Atlantic and perhaps the Baltic, the first Greek to be exposed to the phenomena of the Arctic, and was the originator of tidal theory.
Satyros (early third century BC) explored the western coast of the Red Sea for Ptolemy II.
Seleukos of Seleukeia (*c.*160–120 BC) wrote the first treatise on the tides.
Septimius Flaccus (*c.*AD 80s) established a route from the oasis of Garama to the Aithiopians.
Simonides (early third century BC) lived at Meroë as Ptolemaic agent for several years, and wrote about the region.
Skylax of Karyanda (latter sixth century BC) explored the Indus and sailed back to Persia on the orders of Dareios I.
Strabo of Amaseia (*c.*65 BC – AD 25) wrote a 17-book *Geography*, which discusses the history of geography up to his day, and is the primary source on the topic.
Straton of Lampsakos (early third century BC) was interested in the formative processes of the earth.
C. **Suetonius Paulinus** (mid-first century AD) was the first Roman to reach the west African tropics.
Thales of Miletos (latter seventh century BC), the original Greek natural philosopher, was the first to theorize about the nature of the earth.
Theophanes of Mytilene (first half of the first century BC) was the first to write about the Caucasus.
Tiberius Claudius Nero (42 BC – AD 37), the future Roman emperor, discovered the source of the Danube and explored the Alps.
Timosthenes of Rhodes (early third century BC), admiral of Ptolemy II, wrote *On Harbors*, a technical sailing manual covering the Mediterranean, Black Sea, and Red Sea.

Xanthos of Lydia (fifth century BC) was interested in the changes that the earth had undergone, and observed fossils and oceanic phenomena far from the sea.

Xenophanes of Kolophon (latter sixth century BC) considered the nature and history of the earth.

Xenophon (early fourth century BC) wrote the *Anabasis*, an autobiographical account of the Greek mercenaries in the service of Cyrus the Younger of Persia.

APPENDIX 2

SOME FURTHER NOTES ON MAPPING IN ANTIQUITY

Maps are an essential tool of geography, but the evidence for them in Greek antiquity remains elusive. Although the concept of a map has existed since Sumerian times,[1] there is little physical evidence for any from the Greek period. One must rely primarily on literary descriptions, which can be ambiguous, and thus their study becomes a philological problem.[2] Morover, maps that are not the work of the author of a text became attached in later times to their manuscripts: this is certainly the case with the *Geographical Guide* of Ptolemy. Ephoros' statement about the four ethnic divisions of the extremities of the inhabited world does not seem to imply a map,[3] yet in the sixth century AD Kosmas Indikopleustes had one in his text of Ephoros.[4] The Artemidorus Papyrus includes a portion of a map of the southern Iberian peninsula (as well as other material, including a supposed excerpt from Book 2 of the *Geographoumena* of Artemidoros of Ephesos, written around 100 BC). Yet the authenticity of the papyrus, and thus its map, remains disputed.[5] If it proves genuine, it will be the best surviving example of a Greek map, although it is more of a sketch or draft than a finished product, and remains difficult to interpret.[6]

Literary evidence, then, is the primary source for Greek maps.[7] Greek map-making began with Anaximandros of

Miletos in the early sixth century BC, but there are no particulars about his efforts.[8] The next known map-maker is his fellow citizen, Hekataios, around 500 BC: his map was said to be so much better than its predecessor that it was "a source of wonder," but again it is little known. Aristagoras, the tyrant of Miletos, used a map that may have been drawn by Hekataios. In discussing it, Herodotos preserved the first detailed information about a Greek map. Aristagoras went to Sparta and

> had with him a bronze tablet on which the entire circuit [of the earth] was engraved, all the sea and all the rivers.[9]

Aristagoras then described to the Spartan king, Kleomenes, what was on the map:

> the lands where they live are next to one another: here are the Ionians and here are the Lydians, who live in a good land and have a great amount of silver ... next to the Lydians are the Phrygians, toward the east, the richest in flocks and the richest in produce of all whom I know about. Near the Phrgyians are the Kappadokians, whom we call Syrians.

The account continues across Anatolia and on to the east, and eventually reaches "Sousa, where the Great King lives and where his treasure storehouses are." Were it not for Herodotos' explicit testimony that Aristagoras is talking about a map, the reader would not necessarily know that the description was anything more than an itinerary from Ionia to Persia. There are many similar itineraries in Greek literature that may not have come from a map, and thus determining the origin of such a report is an inherent problem in understanding the evidence for ancient maps.[10] Herodotos also made it clear that maps could provide wrong information, as well as be more revealing than its creator might have intended: the reaction of Kleomenes upon seeing Aristagoras' map was that the Persians were too far away to be a threat.[11]

Other maps, and itineraries that may be based on maps, are documented for classical times but, as always, it is difficult to separate the literary genre from the physical map. Whether or not works such as Demokritos' *Kosmographia* or *The Circuit of the Earth* of Eudoxos of Knidos included maps cannot be proven.[12] Yet it seems that maps were widespread, especially by the early Hellenistic period, as demonstrated by the will of Theophrastos, in which he requested that there be

> erected in the lower stoa [of the Lyceum in Athens] plans on which the circuits of the earth are depicted.[13]

This enigmatic statement cannot easily be interpreted: obviously there was more than one plan, but the use of the plural throughout suggests more than one "circuit of the earth," perhaps indicating that the plans were designed to illustrate specific texts. These maps are clearly a teaching tool, not a political document like that of Aristagoras. Using maps for teaching is obvious, but the evidence is elusive.[14] On the other hand, a map such as that described by Apollonios in the palace of Medea's father, showing "all the roads," suggests a political rather than an academic purpose.[15]

Eratosthenes' account of the grid of the *oikoumene* implies a map, and his terminology points in that direction, but it is impossible to be certain.[16] Nevertheless, he used maps in his research, at times objecting to their inaccuracy.[17] His critic, Hipparchos, a mathematician rather than a geographer, probably did not create a map, despite his astronomical positioning of places in the inhabited world, yet he also made use of them.[18] The globe of Krates has such a specific ideological purpose that it can hardly be called a map, although it was surprisingly influential in medieval times.[19] In the later Hellenistic period presumably there were many maps in circulation, and some of these may be reflected in the itineraries preserved in literature. The major Greek mapping project from the Roman period is that of Ptolemy of Alexandria, dated to the mid-second century AD, based on the work of Marinos of Tyre.[20]

The evidence is stronger for Roman maps.[21] As the Romans established their government in new areas, there was a need to have the visual information that a map would provide as part of the political record. In 174 BC, Ti. Sempronius Gracchus dedicated in Rome a map of Sardinia, which he had just conquered.[22] But, more importantly, the centralized government of the Roman state and the belief that Rome ruled the entire *oikoumene*, or *orbis terrarum*, to use the common Latin term, resulted in the production of large-scale public maps. Roman public mapping is typified by that of M. Vipsanius Agrippa of the latter first century BC, which was described as "the lands of the globe for the city to look at," with "globe" refering to the *oikoumene*, not a sphere.[23] Extensive fragments of one such map, the *Forma Urbis Romae*, dating to around AD 200, actually survive,[24] and another, from maybe a century later, exists through the medium of the document known as the Peutinger Map.[25]

Yet maps did not need to be state documents. One of the more engaging accounts of a Roman map comes from a poem of Propertius, probably written around 16 BC.[26] Arethusa pines for her husband, away on military service, yet a map helps her determine where he is:

> I am forced to learn from a tablet the worlds painted on
> it, and how they were placed on it by its learned creator:
> which lands are slow with frost, which decay from heat,
> and what wind will bring the sails safely to Italy.[27]

In fact, the impression is that Arethusa has both a map and an accompanying geographical commentary. Propertius' vignette demonstrates that maps were far more common in Roman times than the scant physical evidence implies.

NOTES

Introduction

1. All known geographical writers are listed in *EANS*, pp. 999–1002.
2. James S. Romm, *The Edges of the Earth in Ancient Thought* (Princeton 1992).
3. Vitruvius 10.9.

Chapter 1 The Beginnings

1. Jean-Marie Kowalski, *Navigation et géographie dans l'antiquité gréco-romaine* (Paris 2012) 100.
2. Emily Vermeule, *Greece in the Bronze Age* (Chicago 1964) 10–12.
3. Homer, *Odyssey* 5.233–61.
4. The artifact is now in the Fitzwilliam Museum, Cambridge (GR. 18.1963); R. V. Nicholls, "Recent Acquisitions by the Fitzwilliam Museum, Cambridge," *AR* 12 (1965–6) 44–5; Lionel Casson, *Ships and Seamanship in the Ancient World* (Princeton 1971) 30–2.
5. M. Cary and E. H. Warmington, *The Ancient Explorers* (Baltimore 1963) 22–3.
6. Sites named Minoa are on the island of Amorgos, in the Megarid and the southeastern Peloponnesos, and on Sicily.
7. Herodotos 7.170; Diodoros 4.79.
8. Peter Green, *The Argonautika: The Story of Jason and the Quest for the Golden Fleece* (Berkeley 1997) 26–7, 41.
9. Pherekydes (*BNJ* #3) F111a (= Apollodoros, *Biblikotheke* 1.9.19).
10. Homer, *Iliad* 7.467–71, 21.40–1, 23.747.
11. Homer, *Odyssey* 12.69–72; 10.135–7; Hesiod, *Theogony* 956–62.
12. Herodotos 4.85; Eratosthenes, *Geography* F117; the islands are the modern Örektaşı.

13. Hesiod, *Theogony* 340, 961, 992–1002.
14. Pausanias 2.1.1, 2.3.10.
15. Hesiod, *Catalogue* F38.
16. Pomponius Mela 1.108.
17. Strabo 11.2.19; Pliny, *Natural History* 33.52; David Braund, *Georgia in Antiquity* (Oxford 1994) 21–5.
18. Stephanie West, "'The Most Marvellous of All Seas': The Greek Encounter with the Euxine," *G&R* 50 (2003) 151–67; Herodotos 3.93, etc.
19. John Boardman, *The Greeks Overseas: Their Early Colonies and Trade* (new and enlarged edition, London 1980) 254; Braund, *Georgia* 96–103.
20. Strabo 1.2.39.
21. Strabo 11.2.17; Braund, *Georgia* 40–2.
22. Braund, *Georgia* 93.
23. Homer, *Iliad* 2.494–759, 815–77.
24. Georgia L. Irby, "Mapping the World: Greek Initiatives from Homer to Eratosthenes," in *Ancient Perspectives: Maps and their Place in Mesopotamia, Egypt, Greece and Rome* (ed. Richard J. A. Talbert, Chicago 2012) 85–6.
25. R. Hope Simpson and J. F. Lazenby, *The Catalogue of Ships in Homer's Iliad* (Oxford 1970).
26. For a good discussion of this issue, see Richmond Lattimore, *The Iliad of Homer* (Chicago 1951) 24–8.
27. Homer, *Odyssey* 3.188–90.
28. Apollodoros of Athens (*FGrHist* #244) F167 (= Strabo 6.1.3).
29. Boardman, *Greeks Overseas* 165–8.
30. Strabo 6.1.15, 5.2.5.
31. It was a four-day trip: Homer, *Odyssey* 3.180–2.
32. Lykophron, *Alexandra* 592–632; Strabo 5.1.9, 6.3.9.
33. Juba of Mauretania, *Roman Archaeology* F3 (= Pliny, *Natural History* 10.126–7).
34. Strabo 5.1.8.
35. Lord William Taylour, *Mycenaean Pottery in Italy and Adjacent Areas* (Cambridge 1955) 81–137.
36. Frank H. Stubbings, "The Recession of Mycenaean Civilization," *CAH* 2.2 (3rd edn, Cambridge 1975) 356–8.
37. E D. Phillips, "Odysseus In Italy," *JHS* 73 (1953) 53–67.
38. Homer, *Odyssey* 9.39–81.
39. Strabo 7 F18; Homer, *Odyssey* 9.193–215; Herodotos 7.59, 110.
40. Homer, *Odyssey* 9.82–104; Herodotos 4.177; Eratosthenes, *Geography* F105 (= Pliny, *Natural History* 5.41); Strabo 1.2.17, 17.3.17.
41. Bernhard Herzhoff, "Lotus," *BNP* 7 (2005) 822–3.

42. Homer, *Odyssey* 9.105–51.
43. Kallimachos, *Hymn* 3.46–50; Vergil, *Georgics* 4.170–3.
44. Homer, *Odyssey* 10.1–75; Antiochos of Syracuse (*FGrHist* #555) F1 (= Pausanias 10.11.3); Thucydides 3.88; Strabo 6.2.10.
45. Homer, *Odyssey* 10.80–132; Hesiod, *Catalogue* F98.
46. Homer, *Odyssey* 10.133–574.
47. Theophrastos, *Research on Plants* 5.8.3, 9.15.1.
48. Strabo 5.3.6.
49. Homer, *Odyssey* 12.165–200; Strabo 1.2.12–13.
50. Homer, *Odyssey* 12.426–53; Thucydides 4.24.4.
51. Thucydides 6.2.
52. Homer, *Odyssey* 12.447–54; Kallimachos F413 (= Strabo 1.2.37).
53. Homer, *Odyssey* 5.272–7.
54. Homer, *Odyssey* 10.86; Krates (*Cratete di Mallo, I frammenti*, ed. Maria Broggiato, La Spezia 2001) F50.
55. Ruth Scodel, "The Paths of Day and Night," *Ordia Prima* 2 (2003) 83–6.
56. Homer, *Odyssey* 11.14–19.
57. Homer, *Iliad* 18.483–9; see also 1.423, 8.485–6, 18.239–40, 23.205–7.
58. For example, Homer, *Iliad* 14.246; *Odyssey* 11.639, 20.65.
59. Homer, *Iliad* 7.422, 18.399; *Odyssey* 10.511, 19.434, 20.65.
60. On Greek attitudes to the Ocean, see Georgia L. Irby, "Hydrology: Ocean, Rivers, and Other Waterways," in *A Companion to Science, Technology, and Medicine in Ancient Greece and Rome* (ed. Georgia L. Irby, London 2016) 182–90.
61. Diodoros 5.20; J. Oliver Thomson, *History of Ancient Geography* (Cambridge 1948) 40–1; see further *infra*, pp. 23–4.
62. Cary and Warmington, *Ancient Explorers* 75–7.
63. Homer, *Iliad* 1.423, 9.381–4; *Odyssey* 1.22, 4.126–7.
64. *Infra*, pp. 23–4.
65. Homer, *Iliad* 6.288–95, 23.740–7.
66. Homer, *Odyssey* 4.561–9; Paul T. Keyser, "From Myth to Map: The Blessed Isles in the First Century BC," *AncW* 24 (1993) 149–67.
67. Plutarch, *Concerning the Face that Appears on the Globe of the Moon* 26.
68. Strabo 3.2.13, 3.4.3–4; Andrew T. Fear, "Odysseus and Spain," *Prometheus* 18 (1992) 19–26.
69. For example, Homer, *Iliad* 12.13–23; *Odyssey* 5.382–7.
70. 1 Kings 5:1–12, 9:26–8.
71. 2 Chronicles 9:21; Carolina López-Ruiz, "Tarshish and Tartessos Revisited: Textual Problems and Historical Implications," in *Colonial Encounters in Ancient Iberia* (ed. Michael Dietler and Carolina López-Ruiz, Chicago 2009) 255–80.

72. Homer, *Odyssey* 15.415.
73. Casson, *Ships* 56–8.
74. Homer, *Odyssey* 15.415–16.
75. Herodotos 2.44, 6.46–7.
76. Rhys Carpenter, "Phoenicians In the West," *AJA* 62 (1958) 35–53.
77. Velleius 1.2.3; Strabo 1.3.2; Pomponius Mela 3.46; Pliny, *Natural History* 16.216.
78. Michael Dietler, "Colonial Encounters in Iberia and the Western Mediterranean: An Exploratory Framework," in *Colonial Encounters in Ancient Iberia* (ed. Michael Dietler and Carolina López-Ruiz, Chicago 2009) 3–48.
79. Timaios (*BNJ* #566) F60 (= Dionysios of Halikarnassos, *Roman Antiquities* 1.74.1).
80. See the excellent map in Hans Georg Niemeyer *et al.*, "Phoenicians, Poeni," *BNP* 11 (2007) 161–2; D. B. Harden, "The Phoenicians on the West Coast of Africa," *Antiquity* 22 (1948) 141–50.
81. The Greek name for the continent is "Libya" (Homer, *Odyssey* 4.85, 14.295), probably originally an ethnym refering to peoples west of Egypt (e.g. Herodotos 4.181), which in time came to be applied to the entire continent except for Egypt and the Kyrenaika. "Africa" is the Roman term, originally the region around Carthage (Ennius, *Annals* F106; *Satires* F10; Sallust, *Jugurtha* 17.1), and again probably first an ethnym, but which also was expanded to mean the whole continent (Pomponius Mela 1.8). As is often the case, the Greek term "Libya" was widely used in Latin but the Latin term "Africa" was rarely used in Greek.
82. Herodotos 4.42; other reports are derivative.
83. Thomson, *History* 6–9.
84. Duane W. Roller, *Through the Pillars of Herakles: Greco-Roman Exploration of the Atlantic* (London 2006) 41.
85. Strabo 2.3.4–5.
86. *Infra*, pp. 141–2.
87. Poseidonios F49 (= Strabo 2.3.4).
88. *Infra*, p. 59.
89. Roller, *Pillars* 25–6.
90. Herodotos 2.161; Diodoros 1.68.1; Josephus, *Against Apion* 1.143.
91. Strabo 17.3.3.
92. *Infra*, p. 60.
93. Homer, *Iliad* 2.867–9; Ephoros (*BNJ* #70) F127 (= Strabo 14.1.6); Vanessa B. Gorman, *Miletos: The Ornament of Ionia* (Ann Arbor 2001) 14–31.

94. Pliny, *Natural History* 5.112.
95. Boardman, *Greeks Overseas* 240–3.
96. Diogenes Laertios 1.22–7, 37–8; Daniel W. Graham, *The Texts of Early Greek Philosophy* (Cambridge 2010) vol. 1, pp. 17–44.
97. Aristotle, *On the Heavens* 2.13 (294a).
98. Seneca, *Natural Questions* 3.14.1.
99. Aristotle, *Metaphysics* 1.3.5.
100. Diogenes Laertios 2.1–2; see also Eusebios, *Preparation for the Gospel* 10.14.11; Suda, "Anaximandros"; Graham, *Texts*, vol. 1, pp. 45–71.
101. Hippolytos, *Refutation* 1.6.3.
102. Diogenes Laertios 2.1; Duane W. Roller, "Columns in Stone: Anaximandros' Conception of the World," *AntCl* 53 (1989) 185–9; Robert Hahn, *Anaximander and the Architects: The Contributions of Egyptian and Greek Architectural Technologies to the Origins of Greek Philosophy* (Albany 2001) 177–218.
103. Eratosthenes, *Geography* F12 (= Strabo 1.1.11).
104. Agathemeros 1; for his text and translation, with commentary, see Aubrey Diller, "Agathemerus: *Sketch of Geography*," *GRBS* 16 (1975) 59–76.
105. Herodotos 4.36.
106. Homer, *Iliad* 6.169; Odyssey 1.141, 12.67.
107. Herodotos 5.49.
108. Aetius, *Placita* 3.3.1; Seneca, *Natural Questions* 2.18.
109. Seneca, *Natural Questions* 2.17; Diogenes Laertios 2.3; Aetius, *Placita* 2.19.2, 2.22.2, 3.3.2, 3.4.1; Pseudo-Plutarch, *Miscellanies* 3; Graham, *Texts*, vol. 1, pp. 72–94.
110. Aristotle, *Meteorologika* 2.1, 7; Hippolytos, *Refutation* 1.7.6.
111. Diogenes Laertios 8.48; see also 9.21.
112. Plato, *Phaidon* 58 (108b–110b); see also Aristotle, *On the Heavens* 2.13; Paul T. Keyser, "The Geographical Work of Dikaiarchos," in *Dicaearchus of Messana: Text, Translation, and Discussion* (ed. William W. Fortenbaugh and Eckart Schütrumpf, New Brunswick 2001) 362–4. For further on Parmenides, see *infra*, pp. 73–4.
113. Plato, *Phaidon* 46; Eratosthenes, *Geography* F25 (= Strabo 1.4.1); Thomson, *History* 110–12.
114. Diogenes Laertios 9.18–20.
115. Clement of Alexandria, *Miscellanies* 7.22.
116. Aetius, *Placita* 2.24.4. Nevertheless the passage has often been interpreted as refering to an eclipse: see Xenophanes of Colophon, *Fragments* (ed. J. H. Lesher, Toronto 1992) 217.
117. Hippolytos, *Refutation* 1.14.5–6.

Chapter 2 The Expansion of the Greek Geographical Horizon

1. Hesiod, *Theogony* 339–41, 1014. The citation of the Etruscans has been questioned since late antiquity: see E. H. Bunbury, *A History of Ancient Geography* (London 1883) vol. 1, pp. 87–8.
2. David A. Lupher, *Romans in a New World* (Ann Arbor 2003), especially pp. 190–4.
3. A. J. Graham, "The Colonial Expansion of Greece," *CAH* 3.3 (2nd edn 1982) 153–5.
4. Thucydides 1.38, 43.
5. Thucydides 1.15.1; Plato, *Laws* 4.708b, 5.740e.
6. Herodotos 4.151.
7. Strabo 6.3.2.
8. Herodotos 4.147–59.
9. Herodotos 4.152; *infra*, pp. 41–2.
10. Hesiod, *Theogony* 339–40.
11. Boardman, *Greeks Overseas* 240.
12. Herodotos 4.78–80.
13. Graham, "Colonial Expansion" 129.
14. Strabo 7.3.13.
15. Boardman, *Greeks Overseas* 247–50.
16. Herodotos 2.33, 4.49.
17. Boardman, *Greeks Overseas* 165–9.
18. Hekataios (*FGrHist* #1) F90; Strabo 5.1.7–8; Giovanni Colonna, "I Greci di Adria," *RSA* 4 (1974) 1–21.
19. Livy 5.33.8; Dionysios of Halikarnassos, *Roman Antiquities* 1.18.3–5.
20. Grave IV of the Shaft Graves at Mykenai yielded 1,290 beads of Baltic amber: see Vermeule, *Greece in the Bronze Age* 89.
21. Herodotos 1.163.
22. Strabo 4.2.1; Barry Cunliffe, *Facing the Ocean: The Atlantic and its Peoples, 8000 BC–AD 1500* (Oxford 2001) 334–5.
23. Herodotos 1.164–7.
24. The most important was Hemeroskopeion (Strabo 3.4.6), at modern Denia, a name reflecting an ancient temple of Artemis that eventually became one of Diana.
25. Rufus Festus Avienus, *Ora maritima* (ed. J. P. Murphy, Chicago 1977). Avienus may not have had direct access to these early sources, but rather used a composite one of the fourth century BC (John Hind, "Pyrene and the Date of the 'Massaliot Sailing Manual'," *RSA* 2 [1972] 39–52). See also Roller, *Pillars* 9–13 and, further, *infra*, p. 44.
26. Herodotos 4.147–59.

27. Herodotos 2.32; R. C. C. Law, "The Garamantes and Trans-Saharan Enterprise in Classical Times," *Journal of African History* 8 (1967) 181–200.
28. Cary and Warmington, *Ancient Explorers* 218; John Ferguson, "Classical Contacts with West Africa," in *Africa in Classical Antiquity* (ed. L. A. Thompson and John Ferguson, Ibadan 1969) 10–11.
29. Herodotos 4.43; *infra*, p. 59.
30. Homer, *Iliad* 3.6.
31. Luigi Luca Cavalli-Sforza, "Demographic Data," in *African Pygmies* (ed. Luigi Luca Cavalli-Sforza, Orlando 1986) 24.
32. Herodotos 2.178; Strabo 17.1.18; Boardman, *Greeks Overseas* 117–34.
33. Diogenes Laertios 1.27.
34. Boardman, *Greeks Overseas* 38–54.
35. Herodotos 4.13–15.
36. Strabo 1.2.10.
37. J. D. P. Bolton, *Aristeas of Proconnesus* (Oxford 1962).
38. Timothy P. Bridgman, *Hyperboreans: Myth and History in Celtic-Hellenic Contacts* (New York 2005) 27–32.
39. Herodotos 1.201; Ptolemy, *Geographical Guide* 6.16.5.
40. John Ferguson, "China and Rome," *ANRW* 2.9.2 (1978) 581–603.
41. Herodotos 4.152.
42. Herodotos 1.163.
43. Pliny, *Natural History* 7.197.
44. Homer, *Iliad* 11.25, etc.
45. Herodotos 3.115.
46. Strabo 2.5.15, 2.5.30, 3.2.9, 3.5.11.
47. Roller, *Pillars* 12–13.
48. James D. Muhly, "Sources of Tin and the Beginnings of Bronze Metallurgy," *AJA* 89 (1985) 285–8.
49. Seneca, *Natural Questions* 4a.2.22; Roller, *Pillars* 15–19.
50. Aristides 36.85–95. If Euthymenes wrote a report, it was probably the first to mention crocodiles and hippopotami in Greek.
51. *Infra*, pp. 56–60.
52. Herodotos 1.84–6.
53. Ronald Syme, *Anatolica: Studies in Strabo* (ed. Anthony Birley, Oxford 1995) 1–23.
54. Herodotos 5.49–54.
55. Isidoros, *Parthian Stations* 1.
56. Herodotos 1.188–92.
57. Herodotos 1.205–14.
58. Herodotos 2.1.

59. Herodotos 3.17–25, 114.
60. Arrian, *Indika* 1.1–4; *Behistun Inscription* 6.
61. Herodotos 3.98.
62. Herodotos 4.44.
63. *BNJ* #709; Suda, "Skylax"; Aristotle, *Politics* 7.13.1; Athenaios 2.70bc; Philostratos, *Life of Apollonios* 3.47.
64. Herodotos 4.44; Scholia to Pseudo-Skylax, *Periplous* 1.
65. Strabo 14.2.20.
66. The extant text under the name of Skylax, generally known today as Pseudo-Skylax, is a work by an unknown author of the early 330s BC that was produced in the same geographical milieu that also yielded Dikaiarchos and Pytheas (Graham Shipley, *Pseudo-Skylax's Periplous: The Circumnavigation of the Inhabited World* [Exeter 2011] 6–8, 15–18). Sometime before the first century BC it became confused with the treatise of Skylax of Karyanda, which was probably already lost (Strabo 12.4.8, 13.1.4).
67. Herodotos 4.44; Jean-François Salles, "Achaemenid and Hellenistic Trade in the Indian Ocean," in *The Indian Ocean in Antiquity* (ed. Julian Reade, London 1996) 253–7.
68. Hesiod, *Catalogue* F98 (= Strabo 7.3.7).
69. Herodotos 4.59–82.
70. Herodotos 1.203.
71. *FGrHist* #1; Lionel Pearson, *Early Ionian Historians* (Oxford 1939) 25–108.
72. Herodotos 2.143, 5.36; Diodoros 10.25.4; Eratosthenes, *Geography* F1 (= Strabo 1.1.1); Agathemeros 1.1.
73. As is often the case with fragmentary works, the titles are uncertain and are only documented much later (Strabo 12.3.22; Athenaios 2.70b).
74. Herodotos 2.143, 5.36, 125; Dionysios of Halikarnassos, *On Thucydides* 5; Strabo 1.2.6.
75. Eratosthenes, *Geography* F12 (= Strabo 1.1.11); Pearson, *Early Ionian Historians* 31–4.
76. Homer, *Iliad* 1.485; Aeschylus, *Persians* 718.
77. Herodotos 4.45.
78. *Airs, Waters, and Places* 13.
79. Hekataios F303–59.
80. Herodotos 4.42; Diogenes Laertios 8.25–6; Roller, *Pillars* 50–1.
81. Eratosthenes, *Geography* F33 (= Strabo 1.4.7).
82. Agathemeros 1.1; *supra*, p. 28.
83. Herodotos 4.36.
84. Pearson, *Early Ionian Historians* 28–9.
85. Aristotle, *Meteorologika* 3.5.

86. Herodotos 2.34.
87. Hekataios F38–89.
88. See also Herodotos 1.145; Strabo 6.1.4, 15.
89. Hekataios F90.
90. Hekataios F91–108; Strabo 6.2.4, 7.5.8.
91. Hekataios F109–37; Herodotos 6.137.
92. Strabo 7.7.1.
93. Hekataios F138–83.
94. Athenaios 10.447d; Andrew Dalby, *Food in the Ancient World From A to Z* (London 2003) 65.
95. Hekataios F184–92; Herodotos 4.59–82.
96. Pearson, *Early Ionian Historians* 65.
97. Hekataios F196–217; Strabo 12.3.22, 14.1.8.
98. Hekataios F218–80.
99. Hekataios F281–99.
100. Athenaios 2.70b.
101. Hekataios F300–24; Herodotos 2.143; Diodoros 1.37; Arrian, *Anabasis* 5.6.5.
102. Herodotos 2.143.
103. Hekataios F324–57; Herodotos 3.17–26.
104. Hekataios F1–35.
105. Demetrios, *On Style* 12.
106. It is reflected at Herodotos 1.1.
107. Strabo 1.1.1.

Chapter 3 The Spread of Geographical Knowledge and Scholarship in the Classical Period

1. Jerker Blomqvist, *The Date and Origin of the Greek Version of Hanno's Periplus* (Lund 1979); for the Greek text and English translation, see Roller, *Pillars* 129–32.
2. *On Marvellous Things Heard* 37; Herodotos 4.196. The arguments regarding the legitimacy of the voyage are summarized in Roller, *Pillars* 29–43.
3. Hanno, *Periplous* 1.
4. Pseudo-Skylax 112; Strabo 1.3.2.
5. Hanno, *Periplous* 14–15.
6. Pliny, *Natural History* 2.169; Arrian, *Indika* 43.11–12.
7. B. H. Warmington, *Carthage* (revised edition, New York 1969) 69–70.
8. Pliny, *Natural History* 2.169.
9. Avienus, *Ora maritima* 117–19, 380–3, 412–13.
10. Philip Freeman, *Ireland and the Classical World* (Austin 2001) 28–33.

11. Roller, *Pillars* 27–9.
12. Athenaios 2.44e.
13. Strabo 2.3.4.
14. Herodotos 4.43.
15. Eratosthenes, *Geography* F154 (= Strabo 17.1.19); Polybios 3.22.
16. Pseudo-Skylax 112; Shipley, *Pseudo-Skylax's Periplous* 201–9.
17. Pliny, *Natural History* 18.22–3; *infra*, pp. 137–9.
18. Pindar, *Paean* 2; *Pythian* 4; *Olympian* 1–6.
19. On this perennial problem, see Paul W. Wallace, *Strabo's Description of Boiotia: A Commentary* (Heidelberg 1979) 168–72.
20. Pindar, *Olympian* 3.44; *Nemean* 3.21–2; *Isthmian* 4.12.
21. *Homeric Hymn to Dionysos* 29.
22. *Supra*, pp. 40–1; Pindar, *Olympian* 3.16; *Pythian* 10.30; *Isthmian* 6.23; *Paean* 8.63.
23. *Supra*, pp. 19–21.
24. Herodotos 4.13, 32–6.
25. Pindar, *Pythian* 4.
26. Aeschylus, *Agamemnon* 281–316.
27. Aeschylus, *Persians* 480–514, 858–900.
28. Aeschylus, *Prometheus Bound* 707–41, 792–815.
29. Hekataios F203; Herodotos 1.28; Strabo 11.14.5, 14.5.24.
30. Aeschylus, F190–201; Strabo 1.2.27, 4.1.7; Arrian, *Periplous of the Euxine Sea* 19.2.
31. Aeschylus F199 (= Strabo 4.1.7).
32. Xanthos (*FGrHist* #765) F12 (= Eratosthenes, *Geography* F15= Strabo 1.3.4).
33. Hippolytos, *Refutation* 1.14.1–6.
34. Empedokles F91–105 Graham; Diogenes Laertios 8.63.
35. Xanthos F13 (= Strabo 12.8.19, 13.4.11).
36. *Suda*, "Herodotos"; Herodotos 9.73, etc.
37. Herodotos 1.181–5; 2.29, 44; 3.12; 4.81.
38. George Rawlinson, *The History of Herodotus* (New York 1860–2), vol. 1, pp. 7–8.
39. Herodotos 2.3, 44.
40. Herodotos 1.1.
41. Herodotos 2.23, 143, 156; 3.38; 4.44; 6.137.
42. For example, Herodotos 2.3–5.
43. Herodotos 2.33.
44. Herodotos 2.23, 32, 39; 3.97–104; 4.1–142.
45. Diogenes Laertios 9.21.
46. *Suda*, "Charon," "Dionysios the Milesian."
47. Ephoros (*BNJ* #70) F180 (= Athenaios 12.515e).
48. Herodotos 2.12; Xanthos F12 (= Strabo 1.3.4).
49. Herodotos 1.178–216; 2.2–182; 3.17–25; 4.5–82.

50. A good example is Herodotos 2.142, the 11,340 years of Egyptian history.
51. Herodotos 7.152.
52. Herodotos 4.42; *supra*, pp. 36–7.
53. Herodotos 2.33, 4.49.
54. Herodotos 2.19–27.
55. These were quoted by Diodoros (1.39.1–6), who seemed to accept the existence of the mountains but not any effect of them on the flow of the Nile.
56. Herodotos 2.12.
57. Herodotos 4.86; Casson, *Ships* 281–96.
58. Thucydides 2.15, 97; 6.2–6.
59. Thucydides 1.24.1, see also 1.46.1.
60. For Thucydides and geography, see Hans-Joachim Gehrke, "Thukydides und die Geographie," *Geographia antiqua* 18 (2009) 133–43.
61. The toponym "Italy" originated in the south of the peninsula, modern Calabria and Basilicata. Thus Antiochos' work was about this region and somewhat to the north, not the whole extent of what is Italy today (Strabo 6.1.4).
62. Antiochos (*FGrHist* #555) F3, 7–13.
63. Dionysios of Halikarnassos, *Roman Antiquities* 1.73.3–5; Strabo 5.4.3.
64. Diodoros 2.32.4; Lloyd Llewellyn-Jones and James Robson, *Ctesias' History of Persia* (London 2010) 11–18.
65. J. M. Bigwood, "Ctesias As Historian of the Persian Wars," *Phoenix* 32 (1978) 36.
66. Photius, *Bibliotheke* 72 (= Ktesias [*FGrHist* #688] F45); Herodotos 3.98–105.
67. Klaus Karttunen, *India in Early Greek Literature* (Helsinki 1989) 83–4.
68. This may, in part, reflect existing popular culture: shortly before Ktesias went to the Persian court, Aristophanes (*Birds* 1553) had mentioned the Skiapodes.
69. Xenophon, *Anabasis* 1.2.20.
70. Xenophon, *Anabasis* 1.2.23–6.
71. Xenophon, *Anabasis* 1.4.6. Myriandos is at modern Ada Tepe on the Turkish coast.
72. Xenophon, *Anabasis* 1.4.11; Eratosthenes, *Geography* F52 (= Strabo 2.1.39).
73. Xenophon, *Anabasis* 1.8.26–9, 2.2.6.
74. Xenophon, *Anabasis* 4.7.24.
75. Xenophon, *Anabasis* 4.1.8–11, 4.4.18, 4.8.20, 5.4.
76. Herodotos 1.180.

77. Xenophon, *Anabasis* 4.5.1–6.
78. Strabo 1.1.1.
79. Jacques Jouanna, *Hippocrate* 2.2: *Airs-Eaux-Lieux* (Paris 2003) 8–10. The date of the treatise is disputed, and it may be Hellenistic but, if that were so, it would be even more anachronistic.
80. Diogenes Laertios 9.21–3.
81. The theory of zones was also attributed to Thales and the Pythagoreans, but this seems unlikely: Aetius 2.12; Karlhaus Abel, "Zone," *RE Supp.* 14 (1974) 987–1188.
82. Poseidonios F49 (= Strabo 2.2.2); I. G. Kidd, *Posidonius* 2: *The Commentary* (Cambridge 1988) 221–2; Johannes Engels, "Geography and History," in *A Companion to Greek and Roman Historiography* (ed. John Marincola, Malden, Mass. 2011) 541–2.
83. Homer, *Odyssey* 5.270–7.
84. The word is not cited in extant literature until Aristotle, *Meteorologika* 1.6 (343a).
85. Herodotos 2.29.
86. The term "antarctic," however, did not appear in literature until perhaps the first century BC (Geminos 5.16, 28; see also the Aristotelian *On the Cosmos* 2 [392a]).
87. Agathemeros 2.
88. G. W. Bowersock, "The East-West Orientation of Mediterranean Studies and the Meaning of North and South in Antiquity," in *Rethinking the Mediterranean* (ed. W. V. Harris, Oxford 2005) 167–78.
89. Plato, *Phaidon* 58 (109–10); *Timaios* 63a.
90. Anna-Dorothee von den Brinken, "Antipodes," in *Trade, Travel, and Exploration in the Middle Ages* (ed. John Block Friedman *et al.*, New York 2000) 27–9.
91. Horace, *Ode* 1.28.1–2: "terrae... mensorem... Archyta"; on Archytas, see Diogenes Laertios 8.79–80.
92. Diogenes Laertios 8.86–8.
93. Eudoxos F272–374; Agathemeros 2; Plutarch, *On Isis and Osiris* 6; François Lasserre, *Die Fragmente des Eudoxos von Knidos* (Berlin 1966) 96–127; Henry Mendell, "Eudoxos of Knidos," *EANS* 310–13.
94. Strabo 1.1.1.
95. Eudoxos, F277–80, 286–302, 325; Strabo 17.1.29.
96. Eudoxos F276a (= Agathemeros 2).
97. Aristotle, *On the Heavens* 2.14 (298a); *Meteorologika* 2.5 (362b).
98. Aristotle, *On the Heavens* 2.14 (298b); the figure can be compared with Eratosthenes' essentially accurate 252,000 stadia. For the various figures, see Aubrey Diller, "The Ancient Measurements of the Earth," *Isis* 40 (1949) 6–9.

99. Aetius 4.1.7: this actually may be a Hellenistic interpolation into the corpus of Eudoxos: see Thomson, *History* 117–18.
100. Strabo 9.1.2.
101. For example, Eratosthenes, *Geography* F68 (= Strabo 2.1.20).
102. Germaine Aujac, "Les modes de representation du monde habité d'Aristote a Ptolémée," *AFM* 16 (1983) 14–19.
103. Aristotle, *Meteorologika* 1.13; Aristotle, *Meteorologica* (trans. H. D. P. Lee, Cambridge, Mass., 1952) 102–5.
104. Aristotle, *On the Heavens* 2.14 (298a).
105. Aristotle, *Meteorologika* 1.13 (350b); Roller, *Eratosthenes* 218.
106. Aristotle, *Meteorologika* 2.5 (362b) and elsewhere in the treatise.
107. Demosthenes, *On Halonnesos* 35; *On the Crown* 48.
108. Pseudo-Skylax 112; Shipley, *Pseudo-Skylax's Periplous* 203–4.
109. Shipley, *Pseudo-Skylax's Periplous* 6–8.
110. Pseudo-Skylax 112; David W. J. Gill, "Silver Anchors and Cargoes of Oil: Some Observations on Phoenician Trade in the Western Mediterranean," *BSR* 56 (1988) 1–12.
111. Timaios F164 (= Diodoros 5.19–20).
112. Roller, *Pillars* 45–50.
113. Cary and Warmington, *Ancient Explorers* 70–1.
114. Herodotos 2.33.
115. Caesar, *Gallic War* 1.1, etc.; see also Cicero, *Against Piso* 81.
116. Pomponius Mela 3.33; *infra*, pp. 88–9.
117. *FHG* vol. 4, pp. 519–20; Cary and Warmington, *Ancient Explorers* 138.
118. Aristotle, *Meteorologika* 1.13 (350b); Ptolemy, *Geographical Guide* 4.8.3.
119. Aristotle, *History of Animals* 7(8).12.
120. Herodotos 3.107.
121. G. L. Barber, *The Historian Ephorus* (Cambridge 1935) remains the most thorough study of the historian; the commentary by Victor Parker to *BNJ* #70 is useful but somewhat idiosyncratic.
122. Polybios 5.33.2; Diodoros 16.76.5; *Suda*, "Ephippos."
123. Ephoros F42 (= Strabo 7.3.9).
124. Ephoros F43, 44a, 51 (= Stephanos of Byzantion, "Tibarania," "Hydra"; Scholia to Apollonios of Rhodes, *Argonautika* 2.845).
125. Ephoros F30a (= Strabo 1.2.28).
126. Ephoros F42, 131–2 (= Strabo 4.4.6, 7.2.1, 7.3.9).

Chapter 4 Pytheas and Alexander

1. Quintus Curtius 4.4.18.
2. Diodoros 17.75; Plutarch, *Alexander* 44; Arrian, *Anabasis* 7.16.
3. Herodotos 1.203; Aristotle, *Meteorologika* 2.1 (354a).

4. Quintus Curtius 6.4.18–19; Arrian, *Anabasis* 3.29.2, 5.26.1–2, 7.16. By late Hellenistic times a connection between the Caspian and External Ocean was taken for granted (Strabo 2.5.18), an error that was not fully corrected until early modern times (Thomson, *History* 127–9).
5. Arrian, *Anabasis* 5.26.2–3.
6. Dikaiarchos F124 (= Strabo 2.4.1). Roller, *Pillars* 57–91 is a thorough report on Pytheas and his journey, with previous bibliography; the fragments of *On the Ocean* have been collected by Christina Horst Roseman, *Pytheas of Massalia: On the Ocean* (Chicago 1994). The title is known from Geminos 6.9, and Kosmas Indikopleustes, *Christian Topography* 149.
7. Polybios 3.59.7; Strabo 1.4.3–5.
8. Polybios 34.5.7 (= Strabo 2.4.2).
9. Barry Cunliffe, *The Extraordinary Voyage of Pytheas the Greek* (London 2001) 56–52.
10. Strabo 1.4.3, 5; 4.4.1.
11. Strabo 2.5.8–15, 4.2.1, who used the form "Prettanike" when citing Pytheas' report. In the Roman period the Greek version of the name seems to have evolved to the more familiar "Brettanike," which Strabo also cited (Roseman, *Pytheas* 45), the predecessor to the Latin "Britannia." Yet modern editors of Strabo have not always been meticulous about the two variants. "Prettanike" and "Brettanike" are carefully distinguished in Duane W. Roller, *The Geography of Strabo* (Cambridge 2014).
12. Polybios 34.5.2 (= Strabo 2.4.1); Diodoros 5.21–2; Strabo 2.5.8.
13. Pliny, *Natural History* 4.103.
14. Diodoros 5.22; Strabo 2.1.18, 2.5.42; Pliny, *Natural History* 2.186–7.
15. Strabo 4.5.5; Pliny, *Natural History* 2.217–18.
16. Diodoros 5.32.3–4; Rhys Carpenter, *Beyond the Pillars of Herakles* (New York 1966) 171–2.
17. Strabo 1.4.2, 2.5.8; Pliny, *Natural History* 2.186–7. For his latitude calculations, see *infra*, p. 89.
18. The itinerary from the Faeroes to Iceland is in the *Landnámabók*, written around AD 900; see also Roseman, *Pytheas* 107. For an analysis of the issues of the location of Thule, see Roller, *Pillars* 78–87.
19. Polybios 34.5.1–4 (= Strabo 2.4.1–3); Strabo 1.4.2, 4.5.5; Pliny, *Natural History* 4.104; Scholia to Apollonios of Rhodes, *Argonautika* 4.761–5a.
20. Pliny, *Natural History* 4.94, 104.
21. Pomponius Mela 3.33.

22. Reinhard Wenskus, "Pytheas und der Bersteinhandel," in *Untersuchungen zu Handel und Verkehr der vor- und frühgeschlichtlichen Zeit in Mittel- und Nordeuropa* 1 (ed. Klaus Düwel *et al.*, Göttingen 1985) 84–108.
23. Strabo 2.4.1.
24. Strabo 2.1.18.
25. There is also at least one notable error: Massalia was said to be on the same latitude as Byzantion (Strabo 2.1.12), yet is more than 2° to the north.
26. Roseman, *Pytheas* 42–4.
27. Pliny, *Natural History* 2.217; the cubit is the Latin translation of the Greek *pechys*.
28. Aetius 3.17.2; see also Strabo 3.2.11.
29. Geminos 6.9; Pliny, *Natural History* 2.187, 4.104; Kleomedes, *Meteora* 1.4.208–10, 2.80.6–9; Martianus Capella 6.595.
30. Eudoxos of Knidos (F11) had assumed that there was a pole star, but in fact there was none in the fourth century BC; for the astronomy and a chart showing the location of the celestial pole at that time, see Pytheas, *L'Oceano* (ed. Serena Bianchetti, Pisa 1998) 109–10, 170. Today Pytheas' tetragon circles Polaris (α Ursae Minoris).
31. Fridtjof Nansen, *In Northern Mists: Arctic Exploration in Early Times* (tr. Arthur G. Chater, New York 1911) vol. 1, p. 73; it is hardly unexpected that Nansen believed Thule was in Norway.
32. Diodoros 17.17–118; Strabo 15.1.10–2.11 and elsewhere. Strabo, although he wrote a *Deeds of Alexander* (Strabo 2.1.9), now lost, was highly selective in what he recorded about Alexander in his *Geography*, concentrating, as one might expect, on the more remote areas.
33. Plutarch, *Alexander* 76.1; Arrian, *Anabasis* 7.25.1.
34. Strabo 15.1.35; Waldmar Heckel, *Who's Who in the Age of Alexander the Great* (Oxford 2006) 95–9.
35. Strabo 15.1.2.
36. On the geographical aspect of the expedition, see Klaus Geus, "Space and Geography," in *A Companion to the Hellenistic World* (ed. Andrew Erskine, Malden, Mass. 2005) 232–45.
37. Arrian, *Anabasis* 1, Preface. Ptolemy became King Ptolemy I of Egypt; Aristoboulos was a military engineer on the expedition: Heckel, *Who's Who* 46, 235–8.
38. Strabo 15.1.30.
39. Homer, *Iliad* 6.290; *Odyssey* 4.83, 618.
40. Diodoros 17.46–7; Quintus Curtius 4.1.16–23; Justin 11.10.8.
41. Strabo 17.1.6; Arrian, *Anabasis* 3.1.4–5; P. M. Fraser, *Ptolemaic Alexandria* (Oxford 1972) vol. 1, pp. 3–37.

42. Homer, *Odyssey* 4.354–9.
43. Quintus Curtius 4.8.5.
44. To be sure, some of these foundations are questionable and rely on uncertain late antique or medieval traditions. For a list, see P. M. Fraser, *Cities of Alexander the Great* (Oxford 1996) 239–43.
45. Fraser, *Cities* 101.
46. Thapsakos may be the Tipshah of 1 Kings 4:24; see also Michal Gawlikowski, "Thapsacus and Zeugma: The Crossing of the Euphrates in Antiquity," *Iraq* 58 (1996) 123–33.
47. Eratosthenes, *Geography* F52, 80, 83, 84, 87, 94.
48. Herodotos 1.178–200; Eratosthenes, *Geography* F87 (= Strabo 16.1.21).
49. Eratosthenes, *Geography* F48 (= Strabo 11.12.4–5).
50. Strabo 11.12.1; Arrian, *Anabasis* 3.20.3.
51. Pliny, *Natural History* 6.43–4.
52. J. F. Standish, "The Caspian Gates," *G&R* 17 (1970) 17–24; Eratosthenes, *Geography* F37, 48, 52, 55–6, 60, 62–4, 77–80, 83–6, 108.
53. Arrian, *Anabasis* 3.21.10.
54. Herodotos 1.201–10.
55. Arrian, *Anabasis* 4.2.2; for its possible location, see A. B. Bosworth, *A Historical Commentary on Arrian's History of Alexander* (Oxford 1980–), vol. 2, p. 19.
56. Thomson, *History* 126.
57. Arrian, *Anabasis* 4.22.4; for the reasons it was renamed, see *infra*, pp. 102–4.
58. Arrian, *Anabasis* 3.23.1; the mountains are the modern Elburz.
59. Eratosthenes, *Geography* F47, 69 (= Strabo 2.1.1, 15.1.11).
60. Herodotos 3.93, 4.204, 6.9.
61. Arrian, *Anabasis* 3.29.2; Bosworth, *Historical Commentary,* vol. 1, pp. 373–4.
62. It is remotely possible that the Aral Sea was known to Alexander's people, and the belief that the Oxos as well as the Iaxartes drained into the Caspian may have been a confusion of a barely known sea with a more familiar one: J. Lennart Berggren and Alexander Jones, *Ptolemy's Geography: An Annotated Translation of the Theoretical Chapters* (Princeton 2000) 173; J. R. Hamilton, "Alexander and the Aral," *CQ* 21 (1971) 110–11.
63. Arrian, *Anabasis* 3.30.7; Bosworth, *Historical Commentary*, vol. 1, pp. 377–9.
64. Quintus Curtius 9.3.8, 9.4.18.
65. Diodoros 17.93; Quintus Curtius 9.2.2–3; Arrian, *Anabasis* 5.26.
66. Arrian, *Anabasis* 7.7.1–2.
67. Arrian, *Anabasis* 6.1.1–6.

68. Arrian, *Anabasis* 6.17–18.
69. Arrian, *Anabasis* 5.26.2.
70. Strabo 15.2.4.
71. Arrian, *Anabasis* 6.21.3.
72. Aristoboulos (*FGrHist* #139) F49 (= Arrian, *Anabasis* 6.22.4–8); Strabo 15.2.5–7; see also Lionel Pearson, *The Lost Histories of Alexander the Great* (New York 1960) 176–80.
73. Heckel, *Who's Who* 183–4.
74. Nearchos: *FGrHist* #133; Onesikritos: *FGrHist* #134; see also Pearson, *Lost Histories* 83–149.
75. Arrian, *Indika* 34–5.
76. Strabo 15.2.11–13; Arrian, *Indika* 30–1.
77. Onesikritos F28 (= Pliny, *Natural History* 6.96–100); Truesdell S. Brown, *Onesicritus* (Berkeley 1949).
78. *Infra*, pp. 141–2; Pearson, *Lost Histories* 147.
79. Juba II, *On Arabia* F5 (= Pliny, *Natural History* 6.96–106); Arrian, *Indika* 30–5; Duane W. Roller, *Scholarly Kings: The Writings of Juba II of Mauretania, Archelaos of Kappadokia, Herod the Great, and the Emperor Claudius* (Chicago 2004) 130–40.
80. Herodotos 3.8–9, 107–13.
81. Herodotos 4.44.
82. Euripides, *Bacchants* 16.
83. Pliny, *Natural History* 12.62; Plutarch, *Sayings of Kings and Commanders* 179ef; Gus W. Van Beek, "Frankincense and Myrrh in Ancient South Arabia," *JAOS* 78 (1958) 141–51.
84. Arrian, *Anabasis* 5.26.2–3.
85. Arrian, Anabasis 7.20.7; *Indika* 18.3; Heckel, *Who's Who* 43.
86. *BNJ* #711; Strabo 16.3.2; Arrian, *Indika* 18.4.
87. For the location of these sites, see D. T. Potts, *The Arabian Gulf in Antiquity* 2: *From Alexander the Great to the Coming of Islam* (Oxford 1990) 154–96.
88. Arrian, *Anabasis* 7.20.7.
89. Theophrastos, *Plant Explanations* 2.5.5; *Research on Plants* 4.7.3, 5.4.7; Athenaios 3.93bc.
90. Eratosthenes, *Geography* F94 (= Strabo 16.3.2).
91. Arrian, *Anabasis* 7.20.7–10.
92. Arrian, *Indika* 43.6–7.
93. Eratosthenes, *Geography* F95 (= Strabo 16.4.4).
94. Theophrastos, *Research on Plants* 9.4.2–6.
95. Juba, *On Arabia* F1 (= Pliny, *Natural History* 6.136–56).
96. Herodotos 3.97.
97. Aeschylus, *Prometheus Bound* 422, 717–21.
98. Eratosthenes, *Geography* F23 (= Arrian, *Anabasis* 5.3.1–4); Diodoros 17.83.1; Bosworth, *Historical Commentary*, vol. 2, pp. 213–17.

99. Quintus Curtius 7.3.23. For its possible location and history see Fraser, *Cities* 148–51; W. W. Tarn, *The Greeks in Bactria and India* (revised 3rd edn, Chicago 1997) 460–2.
100. It is possible that the indigenous name may have suggested this (Thomson, *History* 126).
101. See further, *infra*, pp. 115–17.
102. Eratosthenes, *Geography* F23 (= Arrian, *Anabasis* 5.3.1–4).
103. Strabo 11.5.5.
104. Pliny, *Natural History* 6.30, 46–9.
105. Ptolemy, *Geographical Guide* 5.9.14, 5.10.4, 6.12.1; see also 6.12.4, 6.18.1.
106. Eratosthenes, *Geography* F24 (= Strabo 11.7.4).
107. *Supra*, pp. 50–1.
108. Arrian, *Anabasis* 3.30.6–9.
109. *FGrHist* #128; Heckel, *Who's Who* 225.
110. Ephoros F30a (= Strabo 1.2.28); Pearson, *Lost Histories* 12–16.
111. Eratosthenes, *Geography* F113 (= Strabo 11.2.15).
112. Strabo 11.7.4.
113. Roller, *Eratosthenes* 140.

Chapter 5 The Legacy of Alexander and Pytheas

1. M. Rostovtzeff, *Social and Economic History of the Hellenistic World* (Oxford 1941) 1035–41.
2. Dikaiarchos F116–27; Keyser, "Geographical Work," 353–72.
3. Agathemeros 2, 5.
4. Eratosthenes, *Geography* F47 (= Strabo 2.1.1–3).
5. Dikaiarchos F124 (= Strabo 2.4.1–3).
6. Aeschylus, *Prometheus Bound* 719–23; Herodotos 1.203, 4.184; but see Aristotle, *Meteorologika* 1.13.
7. The technique is explained in detail by Keyser, "Geographical Work," 354–61.
8. Euclid, *Optics* 18–19.
9. Dikaiarchos F118–19 (= Pliny, *Natural History* 2.162; Geminos 17.5); Keyser, "Geographical Work," 357.
10. Kleomedes 1.7; Philoponos, *Commentary on the Meteorologika of Aristotle* 1.3.
11. Pliny, *Natural History* 2.162.
12. Dikaiarchos F121 (= Martianus Capella 6.590–1).
13. Archimedes, *Sand Reckoner* 1: his own suggestion was 3 million stadia.
14. Roller, *Eratosthenes* 142–3.
15. Kleomedes 1.5; Thomson, *History* 154.
16. Fraser, *Ptolemaic Alexandria*, vol. 1, p. 414.

17. Dikaiarchos F126–7.
18. Diogenes Laertios 5.59–60.
19. Xenophanes F59 Graham; Herodotos 2.12; Strabo 1.3.4–5.
20. Straton F54 (= Strabo 1.3.4).
21. Diodoros 5.47.4–5.
22. Roller, *Eratosthenes* 130–1.
23. Strabo 2.3.5; Diogenes Laertios 5.58; Roller, *Eratosthenes* 11–12.
24. Agatharchides F1, 57; Strabo 16.4.7, 17.1.5; Agatharchides, *On the Erythraean Sea* (trans. and ed. Stanley M. Burstein, London 1989) 4–6; Lionel Casson, "Ptolemy II and the Hunting of African Elephants," *TAPA* 123 (1993) 247–60.
25. L. A. Thompson, "Eastern Africa and the Graeco-Roman World (to A. D. 641)," in *Africa in Classical Antiquity* (ed. L. A. Thompson and J. Ferguson, Ibadan 1969) 26–61.
26. Herodotos 2.29–30; Fraser, *Ptolemaic Alexandria*, vol. 1, pp. 521–2.
27. BNJ #666; Pliny, *Natural History* 6.183, 194.
28. *Infra*, p. 158; Stanley M. Burstein, *Commentary* to *BNJ* #666.
29. *BNJ* #667; Pliny, *Natural History* 5.59, 6.191.
30. Bion (*BNJ* #668) F1; Pliny, *Natural History* 6.177, 180, 191; Athenaios 13.566c.
31. *BNJ* #669; Pliny, *Natural History* 6.183.
32. Hipparchos, *Against the "Geography" of Eratosthenes* F17 (= Strabo 2.1.20); Stanley M. Burstein, *Commentary* to *BNJ* #670.
33. Agatharchides, ed. Burstein, p. 138; Strabo 2.1.20; Pliny, *Natural History* 37.108. An ancient topaz mine of the Ptolemaic era has been discovered on Zabargad Island: see James A. Harrell, "Discovery of the Red Sea Source of *Topazos* (Ancient Gem Periodot) on Zabargad Island, Egypt," in *Twelfth Annual Sikankas Symposium. Periodot and Uncommon Gem Minerals* (ed. Lisbet Thoresen, Fallbrook, Cal. 2014) 16–30.
34. Diodoros 3.42.1.
35. Strabo 16.4.18–19.
36. Strabo 16.4.5.
37. Diodoros 3.18.4.
38. Pliny, *Natural History* 37.24; Aelian, *Research on Animals* 17.8–9; Athenaios 4.183f, 14.634a.
39. Pliny, *Natural History* 5.47; Agathemeros 7; Emil August Wagner, *Die Erdbeschreibung des Timosthenes von Rhodus* (Leipzig 1888); Fraser, *Ptolemaic Alexandria* vol. 1, pp. 522–3.
40. Eratosthenes, *Geography* F134 (= Strabo 2.1.40); Strabo 13.2.5, 17.3.6; Pliny, *Natural History* 6.15, 163, 183; Agathemeros 20.
41. Agatharchides, ed. Burstein; Fraser, *Ptolemaic Alexandria* vol. 1, pp. 539–50.

42. Steven E. Sidebotham, "Ports of the Red Sea and the Arabia-Indian Trade," in *The Eastern Frontier of the Roman Empire* (ed. D. H. French and C. S. Lightfoot, *BAR-IS* 553, 1989) 485–513.
43. For an excellent and thorough account of the city, and the recent excavations at the site, see Steven E. Sidebotham, *Berenike and the Ancient Maritime Spice Route* (Berkeley 2011).
44. Pliny, *Natural History* 6.102–4.
45. Eratosthenes, *Geography* F34, 53, 57, 98.
46. Herodotos 3.111; J. Innes Miller, *The Spice Trade of the Roman Empire* (Oxford 1969) 153–72; Dalby, *Food* 87–8.
47. Strabo 15.1.22, but the passage is unclear.
48. Strabo 2.1.13. Taprobane (modern Sri Lanka) was actually farther south, but was erroneously located: see *infra*, pp. 118–19.
49. Eratosthenes F34 (= Strabo 2.5.7).
50. For other even more obscure Ptolemaic geographical writers, see Fraser, *Ptolemaic Alexandria* vol. 1, pp. 520–5.
51. Heckel, *Who's Who* 246–8.
52. *BNJ* #428; Pliny, *Natural History* 6.49; John R. Gardiner-Garden, "Greek Conceptions on Inner Asian Geography and Ethnography from Ephoros to Eratosthenes," *Papers On Inner Asia* 9 (Bloomington, Ind. 1987) 44–8.
53. Herodotos 1.201.
54. For an analysis of the list, see Pliny, *Histoire Naturelle* 6, part 2 (ed. Jacques André and Jean Filliozat, Paris 2003) 66–9.
55. *BNJ* #712; Diodoros 19.100.5–6; Plutarch, *Demetrios* 47.3; Gardiner-Garden, "Greek Conceptions," 39–44.
56. Arrian, *Anabasis* 7.16.3.
57. Strabo 2.1.6, 17; 11.7.1, 3; 11.11.6. The area of the Caspian is about 15 per cent less than that of the Black Sea.
58. For example, Strabo 11.6.1.
59. Pomponius Mela 3.38; Pliny, *Natural History* 6.36.
60. The direction of flow, from the Ocean to the Caspian, need not be a problem, as Euthymenes (*supra*, pp. 44–5) had suggested many years previously that the source of the Nile was an inflow from the Atlantic.
61. Ptolemy, *Geographical Guide* 5.9.12–21, 6.14.1–9.
62. Strabo 2.1.15, 11.7.3, 11.11.5. Patrokles' use of Persian parasangs suggests that some of his information was gathered from existing Persian reports.
63. Eratosthenes, *Geography* F69, 73 (= Strabo 2.1.7, 15.1.11).
64. Hipparchos, *Against the "Geography" of Eratosthenes* F12 (= Strabo 2.1.4).
65. Klaus Karttunen, *India and the Hellenistic World* (Helsinki 1997) 257–64.

66. Plutarch, *Alexander* 62.4.
67. Strabo 15.2.9; Justin 15.4.12–21; Appian, *Syriaka* 55.
68. *BNJ* #715; Strabo 2.1.9; Arrian, *Anabasis* 5.6.2.
69. Heckel, *Who's Who* 248–9; A. B. Bosworth, "The Historical Setting of Megasthenes' *Indica*," *CP* 91 (1996) 113–27.
70. Strabo 15.1.11.
71. Josephus, *Jewish Antiquities* 10.227.
72. Karttunen, *India in Early Greek Literature* 96–9.
73. Strabo 15.1.58–60.
74. Strabo 15.1.36, 50–1; Arrian, *Indika* 10.5.7.
75. Diodoros 2.35.1–2; Strabo 15.1.11–12.
76. Strabo 2.1.19–20, 15.1.13; Pliny, *Natural History* 6.69; Arrian, *Indika* 4.2, 5.1.
77. Strabo 15.1.37.
78. Strabo 2.1.9, 15.1.56–7.
79. Pliny, *Natural History* 6.81.
80. Eratosthenes, *Geography* F74 (= Strabo 15.1.14).
81. Strabo 15.1.37; Ferguson, "China and Rome," 582–5.
82. On the development of the silk trade, see Manfred G. Raschke, "New Studies in Roman Commerce with the East," *ANRW* 2.9 (1978) 606–50.
83. *BNJ* #716; Strabo 2.1.9, 14, 17; 15.1.12; Athenaios 9.394e.
84. Eratosthenes, *Geography* F22, 67 (= Strabo 2.1.9, 19).
85. Karttunen, *India in Early Greek Literature* 100–1.
86. Karttunen, *India and the Hellenistic World* 264–71.
87. Roller, *Pillars* 50–1.
88. Eratosthenes, *Geography* F39 (= Strabo 1.1.18).

Chapter 6 Eratosthenes and the Invention of the Discipline of Geography

1. Roller, *Eratosthenes* 7–15; Klaus Geus, *Eratosthenes von Kyrene* (Munich 2002).
2. Suetonius, *Grammarians* 10.
3. Roller, *Eratosthenes* 263–7.
4. Heron of Alexandria, *Dioptra* 35; Galen, *Institutes of Logic* 26–7; Macrobius, *Commentary on the Dream of Scipio* 1.20.9.
5. Aristotle, *On the Heavens* 2.14 (298b).
6. The best ancient discussion of his technique is Kleomedes 1.7; see also Geminos 16.6–9; Macrobius, *Commentary on the Dream of Scipio* 2.6.2–5; Vitruvius 1.6.9; and Bernard R. Goldstein, "Eratosthenes on the 'Measurement' of the Earth," *Historica Mathematica* 11 (1984) 411–16.
7. Martianus Capella 6.596–8.

8. Strabo 2.1.20.
9. There was also the story of a well at Syene whose bottom was illuminated by the sun at the solstice, but this is not documented until Roman times, when it became a tourist attraction (Pliny, *Natural History* 2.183).
10. Fraser, *Ptolemaic Alexandria*, vol. 1, pp. 414–15.
11. James Evans and J. Lennart Berggren, *Geminos's Introduction to the Phenomena* (Princeton 2006) 211–12.
12. Aubrey Diller, "Geographical Latitudes in Eratosthenes, Hipparchus and Posidonius," *Klio* 27 (1934) 258–69.
13. Roller, *Eratosthenes* 271–3; Thomson, *History* 161–2.
14. Pliny, *Natural History* 2.247.
15. Arrian, *Anabasis* 5.3.1–4, 5.6.2–3; *Indika* 3.1–5.
16. Strabo 2.1.1, 2.5.24, 8.8.4.
17. O. A. W. Dilke, *Greek and Roman Maps* (Ithaca 1985) 33–5; Thomson, *History* 135, 142; Irby, "Mapping the World," 101–4.
18. Eratosthenes, *Geography* F15, 25 (= Strabo 1.3.3–4, 1.4.1).
19. Eratosthenes, *Geography* F33 (= Strabo 1.4.6).
20. Eratosthenes, *Geography* F30 (= Strabo 2.5.5–6).
21. On the unusual geographical diction of Eratosthenes, see Klaus Geus, "Measuring the Earth and the *Oikoumene*: Zones, Meridians, *Sphragides* and Some Other Geographical Terms Used by Eratosthenes of Cyrene," in *Space in the Roman World* (*Antike Kultur und Geschichte* 5, ed. Kai Brodersen, Münster 2004) 11–26.
22. Eratosthenes, *Geography* F30–1 (= Strabo 2.5.5–6, 13).
23. Eratosthenes, *Geography* F40–3 (= Strabo 2.1.20); Pliny, *Natural History* 2.183–5, 6.171; Ammianus Marcellinus 22.15.31.
24. Eratosthenes, *Geography* F47 (= Strabo 2.1.1–3).
25. Eratosthenes, *Geography* F131 (= Strabo 2.1.41).
26. Eratosthenes, *Geography* F56 (= Strabo 2.1.33).
27. Eratosthenes, *Geography* F34 (= Strabo 2.5.7–9).
28. Eratosthenes, *Geography* F37 (= Strabo 1.4.5).
29. Eratosthenes, *Geography* F66 (= Strabo 2.1.22); Roller, *Eratosthenes* 26.
30. Serena Bianchetti, "Il valore del racconto di viaggio nell'opera geografica di Eratosthene," in *Vermessing der Oikumene* (ed. Klaus Geus and Michael Rathmann, Berlin 2013) 77–86; Katherine Clarke, *Between Geography and History: Hellenistic Constructions of the Roman World* (Oxford 1999) 207.
31. Roller, *Eratosthenes* 164–5.
32. Eratosthenes, *Geography* F92 (= Strabo 2.1.32).
33. The toponyms are listed in Roller, *Eratosthenes* 223–48.
34. Eratosthenes, *Geography* F108 (= Strabo 8.8.9).

35. Eratosthenes, *Geography* F110 (= Strabo 11.6.1).
36. Eratosthenes, *Geography* F98, 107 (= Strabo 17.1.2, 17.3.8.
37. Strabo 2.1.41.
38. Eratosthenes, *Geography* F60 (= Strabo 2.5.40).
39. Eratosthenes, *Geography* F155 (= Strabo 1.4.9).
40. Mention in this passage of the Romans and the excellence of their government, which Eratosthenes hardly knew about, is probably a gloss, even an unconscious one, by Strabo.
41. D. R. Dicks, *The Geographical Fragments of Hipparchus* (London 1960) 3–6.
42. Hipparchos, *Against the "Geography" of Eratosthenes* F34 (= Strabo 2.1.41).
43. Hipparchos, *Against the "Geography" of Eratosthenes* F12–15 (= Strabo 2.1.1–11).
44. Dicks, *Geographical Fragments* 32–3.
45. Hipparchos, *Against the "Geography" of Eratosthenes* F50 (= Strabo 2.5.39).
46. Hipparchos, *Against the "Geography" of Eratosthenes* F12–14 (= Strabo 2.1.4, 7, 11), F53–5 (= Strabo 1.4.4, 2.1.12, 2.5.8).
47. Hipparchos, *Against the "Geography" of Eratosthenes* F39, 43–61.
48. Hipparchos, *Against the "Geography" of Eratosthenes* F48 (= Strabo 2.5.38).
49. Hipparchos, *Against the "Geography" of Eratosthenes* F11 (= Strabo 1.1.12).
50. Eratosthenes, *Geography* F1 (= Strabo 1.1.1); Francesco Prontera, *Geografia e storia nella Grecia antica* (Florence 2011) 3–14.
51. Lionel Pearson, *The Local Historians of Attica* (Atlanta 1981) 31–2.
52. Suetonius, *Grammarians* 2; Strabo 1.1.7, 1.2.24, 31.
53. Strabo 1.2.31; see also Geminos 6.10–21.
54. Strabo 2.5.10. A drawing of the globe appears in Thomson, *History* 203.
55. Dilke, *Greek and Roman Maps* 36–7.
56. Tomislav Bilić, "Crates of Mallos and Pytheas of Massalia: Examples of Homeric Exegesis in Terms of Mathematical Geography," *TAPA* 142 (2012) 295–328.
57. *FGrHist* #2013; Strabo 13.1.45.
58. Strabo 1.2.38.
59. *FGrHist* #244; Strabo 8.3.6.
60. Strabo 13.1.36.
61. Lawrence Kim, "The Portrait of Homer in Strabo's *Geography*," *CP* 102 (2007) 363–8.
62. Bunbury, *History*, vol. 2, pp. 61–9; R. Stiehle, "Der Geograph Artemidoros von Ephesos," *Philologus* 11 (1856) 193–244; Johannes Engels, "Artemidoros of Ephesos and Strabo of Amasia,"

in *Intorno al Papiro di Artemidoro* 2: *Geografia e Cartografia* (ed. C. Gallazzi *et al.*, Rome 2012) 139–55.
63. *Infra*, p. 213.
64. Strabo 15.1.72.
65. See Bunbury, *History*, vol. 2, pp. 65–6, for the calculations.
66. Strabo 1.1.8–9, 3.5.9; Duane W. Roller, "Seleukos of Seleukeia," *AntCl* 74 (2005) 111–18.
67. Plutarch, *Platonic Questions* 8.1.

Chapter 7 The New Roman World

1. Pliny, *Natural History* 18.22–3.
2. Polybios 3.22; Strabo 17.1.19.
3. The earliest Greek reference to Rome, if authentic, seems to be by Damastes of Sigeion (Dionysios of Halikarnassos, *Roman Antiquities* 1.72). Aristotle certainly knew about the Romans (Plutarch, *Camillus* 22.3).
4. Boardman, *Greeks Overseas* 228–9; see also, for the first Greek literary reference to the Etruscans (although not undisputed), Hekataios F59 (= Stephanos of Byzantion, "Aithale: Tyrsenian island").
5. Diodoros 14.93.3–4; Appian, *Italika* 8.1.
6. Robert K. Sherk, "Roman Geographical Exploration and Military Maps," *ANRW* 2.1 (1974) 534–62; Susan P. Mattern, *Rome and the Enemy: Imperial Strategy in the Principate* (Berkeley 1999) 24–80.
7. Polybios 3.48.12; F. W. Walbank, *Polybius* (Berkeley 1972) 11.
8. Nevertheless, Herodotos (4.49) thought "Alpis" was the name of a river.
9. Eratosthenes, *Geography* F51, 147 (= Strabo 2.1.11); Stephanos of Byzantion, "Tauriskoi."
10. Polybios 34.15.7; Marijean H. Eichel and Joan Markley Todd, "A Note on Polybius' Voyage to Africa in 146 BC," *CP* 71 (1976) 237–43.
11. Polybios 34.15.9; Strabo 1.3.2.
12. Pliny, *Natural History* 6.199–201.
13. Roller, *Pillars* 100–4.
14. Geminos 16.32.
15. Eratosthenes, *Geography* F45 (= Strabo 2.3.2); see also Strabo 2.5.7.
16. Polybios 34.1.7–18.
17. Polybios 34.10.6–7.
18. Strabo 4.2.1.
19. Polybios 12.28.1, 3.59.8.
20. The geographical section survives today only in quotation, mostly from Strabo and Pliny, which have been collected as Book

34 of the *Histories*; see also Strabo 8.1.1; F. W. Walbank, "The Geography of Polybius," *ClMed* 9 (1947) 155–82.
21. Pausanias 8.30.8.
22. Strabo 17.3.2.
23. Polybios 3.95.5, 10.10.1–2; Livy 22.19.5.
24. Cicero, *Letters to Friends* 5.12.2.
25. Strabo 3.3.1. His data from Brutus' report may have been obtained derivatively from Polybios or Poseidonios.
26. *Periplous of the Erythraian Sea* 26; Lionel Casson, *The Periplus Maris Erythraei* (Princeton 1989) 159.
27. Poseidonios F49 (= Strabo 2.3.4–5); there are derivative comments by Pomponius Mela (3.90) and Pliny (*Natural History* 2.169); see also J. H. Thiel, *Eudoxus of Cyzicus* (Groningen 1939); Kidd, *Commentary* 240–57; Roller, *Pillars* 107–11.
28. Aristotle, *Meteorologika* 2.5.
29. Thiel, *Eudoxus* 18; Casson, *Periplus* 224; James Beresford, *The Ancient Sailing Season* (Leiden 2013) 213–35.
30. *Periplous of the Erythraian Sea* 1; Casson, *Periplus* 13–14, 96.
31. Strabo 2.5.12.
32. Strabo 1.2.1.
33. *BNJ* #184; Strabo 11.5.1, 13.1.55; Pliny, *Natural History* 8.36, 37.61; see also Scholia to Apollonios, *Argonautika* 4.834.
34. Plutarch, *Lucullus* 23–32; Appian, *Mithridateios* 89–90.
35. Eratosthenes, *Geography* F87 (= Strabo 16.1.21–2).
36. The fragments of his writings are collected as *BNJ* #188.
37. Plutarch, *Pompeius* 33–6.
38. Theophanes F5 (= Strabo 11.14.4).
39. Theophanes F6 (= Strabo 11.14.1); see also Strabo 11.4.1–8.
40. Jacqueline Fabre-Serris, "Comment parler des Amazones? L'exemple de Diodore de Sicile et de Strabon," *CRIPEL* 27 (2008) 46–8.
41. Appian, *Mithridateios* 103.
42. H. H. Scullard, *From the Gracchi to Nero* (4th edn, London 1976) 425.
43. Plutarch, *Crassus* 16.2.
44. I. G. Kidd, *Posidonius* 3: *The Translation of the Fragments* (Cambridge 1999) 3–5.
45. Poseidonios F49 (= Strabo 2.2.1–3.8); Kidd, *Commentary* 216–75; Clarke, *Between Geography and History* 139–54.
46. Strabo 1.1.1.
47. Aristotle, *Meteorologika* 2.5.
48. Seneca, *Natural Questions* 1, Preface 13.
49. For example, Poseidonios F269 (= Strabo 3.4.17).
50. Strabo 2.4.2.

51. J. J. Tierney, "The Celtic Ethnography of Posidonius," *Proceedings of the Royal Irish Academy* 60C5 (1960) 189–275.
52. Strabo 4.4.2–6; Athenaios 13.594de, 4.151e-152f, 154ac, 6.246cd (= Poseidonios F66–9).
53. Diodoros 5.25–32; Caesar, *Gallic War* 6.11–28.
54. Poseidonios F227 (= Strabo 5.2.11).
55. Poseidonios F202, 204 (= Kleomedes 1.7; Strabo 2.5.14).
56. Poseidonios F217–18 (= Strabo 1.1.7, 3.5.7–8).
57. Kleomedes 2.1, 3.
58. Poseidonios F221, 223, 246 (= Strabo 1.3.9, 3.5.5–6, 17.3.10).
59. Strabo 16.2.10.
60. Leandro Polverini, "Cesare e la geografia," *Semanas de estudios romanos* 14 (2005) 59–72.
61. Caesar, *Gallic War* 6.24; Kidd, *Commentary* 308.
62. Caesar, *Gallic War* 1.37–43.
63. Poseidonios F219. The river is also named in the Aristotelian *On Marvellous Things Heard* (168), whose date is uncertain.
64. Caesar, *Gallic War* 3.9, 21; 4.10; 6.24.
65. Caesar, *Gallic War* 4.20–36, 5.7–23.
66. Caesar, *Gallic War* 4.20.
67. Caesar, *Gallic War* 5.13. The difference between the length of night at the summer solstice in London and Paris is 28 minutes.
68. Plutarch, *Caesar* 23.2.
69. Lucan 10.268–331; Suetonius, *Divine Julius* 52.1; Appian, *Civil War* 2.90.
70. Eratosthenes, *Geography* F34 (= Strabo 2.5.7).
71. Strabo 7.3.17.
72. Eratosthenes, *Geography* F34, 53, 58 (= Strabo 2.5.7, 2.5.14, 2.2.2).
73. Eratosthenes, *Geography* F30 (= Strabo 2.5.6).
74. Eratosthenes, *Geography* F33 (= Strabo 1.4.6).
75. Eratosthenes, *Geography* F30 (= Strabo 2.5.5).
76. Cicero, *Republic* 6.20–2; Pliny, *Natural History* 2.171–5.
77. Seneca, *Medea* 375–9.
78. Plutarch, *Concerning the Face that Appears on the Globe of the Moon* 26; Lucian, *True Narrative* 1.
79. Roller, *Pillars* 54–6.
80. R. Scheckley, "Romans in Rio?" *Omni* 5 (June 1983) 43; Frank J. Frost, "Voyages of the Imagination," *Archaeology* 46.2 (March/April 1993) 47.
81. Casson, *Periplus* 123.
82. Dalby, *Food* 159.
83. *Periplous of the Erythraian Sea* 64; see also Casson, *Periplus* 238–41.
84. Lucan 10.142.
85. Thomson, *History* 178.

Chapter 8 Geography in the Augustan Period

1. Strabo 17.1.24–54; Pliny, *Natural History* 6.181–2; Dio 54.5.4–6.
2. For a complete study, see Duane W. Roller, *The World of Juba II and Kleopatra Selene: Royal Scholarship on Rome's African Frontier* (London 2003); for the fragments of Juba's writings, Roller, *Scholarly Kings* 1–169, and *BNJ* #275.
3. Pliny, *Natural History* 5.51–4.
4. Vitruvius 8.2.8.
5. Strabo 15.1.13.
6. Vitruvius 8.2.6–7; Strabo 17.3.4; Pausanias 1.33.6; Dio 76.13.3.
7. Ptolemy, *Geographical Guide* 1.9, 17; also 4.6–8.
8. John Hanning Speke, *Journal of the Discovery of the Source of the Nile* (New York 1868) 420–1; Thomson, *History* 268–9, 275–7.
9. Pliny, *Natural History* 6.202–5; Herodotos 4.49; Roller, *Pillars* 47–9.
10. Pliny, *Natural History* 6.175–80.
11. Velleius 2.101–2; Pliny, *Natural History* 6.141, 12.55–6, 32.10; see also Roller, *Juba* 212–26.
12. G. W. Bowersock, "Perfumes and Power," in *Profumi d'Arabia* (ed. Alessandra Avanzini, Rome 1997) 543–56.
13. Strabo 16.4.22–4; G. W. Bowersock, *Roman Arabia* (Cambridge, MA, 1983) 46–9. On whether the expedition was a success or failure, see Bowersock, "Perfumes and Power," 551–3.
14. Dario Nappo, "On the Location of Leuke Kome," *JRA* 23 (2010) 335–48.
15. Pliny, *Natural History* 6.136–56, 165–7.
16. *BNJ* #781; Pliny, *Natural History* 6.141 (erroneously called "Dionysios," not "Isidoros"); Roller, *Juba* 217–19; Wilfrid H. Schoff, *Parthian Stations of Isidore of Charax* (London 1914).
17. Pliny, *Natural History* 6.145.
18. On the routes involved, see Fergus Millar, "Caravan Cities: the Roman Near East and Long-Distance Trade by Land," in *Essays in Honor of Geoffrey Rickman* (ed. Michael M. Austin *et al.*, London 1998) 123–6.
19. Pliny, *Natural History* 6.96–106, 12.51–81.
20. Pliny, *Natural History* 6.175.
21. Strabo 2.5.12.
22. Eratosthenes, *Geography* F150 (= Caesar, *Gallic War* 6.24).
23. Barbara Levick, *Tiberius the Politician* (London 1976) 27–8.
24. Strabo 7.1.5; Pomponius Mela 3.24; Pliny, *Natural History* 9.63.
25. "Danuvius" is the most common form of the name in Latin; Greek sources usually continued to use "Istros."
26. The Nile had different names in its upper course, and this is the case today with both the Ohio and Missouri.
27. Caesar, *Gallic War* 6.25; Diodoros 5.25.

28. Strabo 7.3.13.
29. Caesar, *Gallic War* 4.17–18, 6.9; Cary and Warmington, *Ancient Explorers* 276.
30. Strabo 7.1.3; Velleius 2.97; Dio 55.1.
31. Livy, *Summary* 141–2; Dio 54.33.1–3.
32. Pomponius Mela 3.33.
33. Velleius 2.97; Dio 54.20.4–6.
34. Olwen Brogan, "Trade Between the Roman Empire and the Free Germans," *JRS* 26 (1936) 195–222.
35. Velleius 2.118–19; Suetonius, *Augustus* 23; Tacitus, *Annals* 1.61; Dio 56.18–22.
36. Tacitus, *Germania* 41.
37. Pliny, *Natural History* 4.96; Ptolemy, *Geographical Guide* 2.11.35.
38. For Archelaos' writings, see *FGrHist* #123; Roller, *Scholarly Kings* 170–6.
39. Diogenes Laertios 2.17; Strabo 1.1.16.
40. Pliny, *Natural History* 37.46; Solinus 52.18–23; see also Cary and Warmington, *Ancient Explorers* 195.
41. Examples include Horace, *Odes* 1.12.56, 3.29.27, 4.15.23; and Ovid, *Amores* 1.14.6.
42. Vergil, *Georgics* 1.121.
43. Strabo 15.1.37.
44. Pausanias 6.26.6–9; see J. G. Frazer, *Pausanias's Description of Greece* (reprint New York 1965) vol. 4, pp. 110–11. Pausanias was uncertain where the Seres lived, believing that Seria was an island in the Erythraian Sea, or somewhere else south or east of Egypt – another example of confusing points along a route with the place of origin of the commodities traded. Much earlier, Aristotle (*History of Animals* 5.19 [551b]) described some indigenous Greek way of weaving the fibers of a cocoon, which probably has nothing to do with Chinese silk. See also Pliny, *Natural History* 6.54.
45. Strabo 2.5.12.
46. Augustus, *Res gestae* 31.1; the choice of words is significant, "Romanorum ducem" indicating that there might have been previous contact with private citizens.
47. Orosios 6.21.
48. Nikolaos (*FGrHist* #90) F100 (= Strabo 15.1.73).
49. Grant Parker, *The Making of Roman India* (Cambridge 2008) 210–17.
50. The itineraries and dates for these journeys are listed in Helmut Halfmann, *Itinera principum* (Stuttgart 1986) 163.
51. Pliny, *Natural History* 3.17.
52. Dilke, *Greek and Roman Maps* 41–53; Pascal Arnaud, "Texte et carte de Marcus Agrippa: historiographie et données textuelles,"

Geographia antiqua 16–17 (2007–8) 73–126; Claude Nicolet, *Space, Geography, and Politics* (Ann Arbor 1991) 95–122; Richard J. A. Talbert, "*Urbs Roma* to *Orbis Romanus*: Roman Mapping on the Grand Scale," in *Ancient Perspectives: Maps and their Place in Mesopotamia, Egypt, Greece and Rome* (ed. Richard J. A. Talbert, Chicago 2012) 167–70; James J. Tierney, "The Map of Agrippa," *Proceedings of the Royal Irish Academy* 63C (1963) 151–66.

53. Pliny, *Natural History* 4.77, 81, 98, 102; 5.9–10; 6.37–9, 57, 196.
54. Dilke, *Greek and Roman Maps* 40.
55. A good example is at *Natural History* 5.73: "below them [the Essenes] was the town of Engada."
56. Strabo 1.1.16; 5.2.7,8; 6.1.11, 6.2.11, 6.3.10.
57. Arnaud, "Texte et carte," 89–94.
58. Vitruvius 8.2.6.
59. For a complete analysis of Strabo and his work, see Roller, *Geography of Strabo* 1–27; Clarke, *Between Geography and History* 193–336.
60. Lawrence Kim, *Homer Between History and Fiction in Imperial Greek Literature* (Cambridge 2010) 47–150.
61. Richard D. Sullivan, "Dynasts in Pontus," *ANRW* 7.2 (1980) 920–2.
62. Johannes Engels, "Kulturgeographie im Hellenismus: Die Rezeption des Eratosthenes und Poseidonios durch Strabon in den *Geographika*," in *Vermessing der Oikumene* (ed. Klaus Geus and Michael Rathmann, Berlin 2013) 87–99.
63. François Lasserre, "Strabon devant l'Empire romain," *ANRW* 30 (1982–3) 867–96.
64. Strabo 1.1.1.

Chapter 9 The Remainder of the First Century AD

1. Pliny, *Natural History* 2.167, 4.97.
2. Velleius 2.106; Dio 55.28.5.
3. Pomponius Mela 3.30–2; Pliny, *Natural History* 4.96–8.
4. Augustus, *Res gestae* 26.
5. Pliny, *Natural History* 2.167.
6. Strabo 2.3.6.
7. Pliny, *Natural History* 4.95; 37.33,36; Ptolemy, *Geographical Guide* 1.11.8.
8. Pliny, *Natural History* 37.45.
9. Tacitus, *Annals* 2.23–6.
10. Seneca the Elder, *Suasoriae* 1.15.
11. Tacitus, *Germania* 28, 37, 41, 46.
12. Casson, *Periplus* 16–18.
13. Pliny, *Natural History* 6.101.
14. See Casson, *Periplus*, for a text, translation, and commentary.

15. The date of this settlement and its temple is uncertain. It is documented on the late-antique map, the Peutinger Map; see also Casson, *Periplus* 24. For the location of Muziris, see Rajan Gurukkhal and Dick Whittaker, "In Search of Muziris," *JRA* 14 (2001) 335–50.
16. Casson, *Periplus* 228–9.
17. *Periplous of the Erythraian Sea* 63; see also Pomponius Mela 3.70; Pliny, *Natural History* 6.80.
18. Casson, *Periplus* 235–6.
19. *Periplous of the Erythraian* Sea 64–5; Dalby, *Food* 206; Casson, *Periplus* 241–3.
20. *Periplous of the Erythraian Sea* 66.
21. Strabo 15.1.15; *Periplous of the Erythraian Sea* 61. There may be a lacuna in the text of the *Periplous* at this point.
22. Pliny, *Natural History* 6.81–91.
23. Samuel Lieberman, "Who Were Pliny's Blue-Eyed Chinese?" *CP* 52 (1957) 174–7.
24. Dio 68.15.
25. *Infra*, p. 199.
26. Ptolemy, *Geographical Guide* 1.13–14.
27. Ptolemy, *Geographical Guide* 7.2; Berggren and Jones, *Ptolemy's Geography* 155–6.
28. Mortimer Wheeler, *Rome Beyond the Imperial Frontiers* (London 1955) 203–10.
29. Berggren and Jones, *Ptolemy's Geography* 173–4.
30. Ptolemy, *Geographical Guide* 7.2.29, 8.27.10; see also Pomponius Mela 3.70.
31. Pliny, *Natural History* 5.11–15; Suetonius, *Gaius* 35; Dio 60.9.
32. F. de la Chapelle, "L'expédition de Suetonius Paulinus dans le sud-est du Maroc," *Hespéris* 19 (1934) 107–24.
33. Pliny, *Natural History* 5.7.
34. P. Salama, "The Sahara in Classical Antiquity," in *General History of Africa* 2: *Ancient Civilizations of Africa* (ed. G. Mokhtar, Paris and London 1981) 513–32.
35. Pliny, *Natural History* 5.36; Law, "Garamantes," 181–200.
36. Ptolemy, *Geographical Guide* 1.8; Berggren and Jones, *Ptolemy's Geography* 145–7.
37. Ptolemy, *Geographical Guide* 1.7–12, 4.6.3, 4.8.5, 7.5.2; Ptolemaios, *Handbuch der Geographie* (ed. Alfred Stückelberger and Gerd Grasshoff, Basel 2006), vol. 1, p. 445.
38. Berggren and Jones, *Ptolemy's Geography* 67; Pliny, *Natural History* 8.70–1.
39. Lucan 10.191–2.
40. Strabo 17.1.5.

41. Seneca, *Natural Questions* 4a.2.3–5; Pliny, *Natural History* 5.55–8.
42. Pliny, *Natural History* 6.181–7.
43. Seneca, *Natural Questions* 6.8.4; see also 6.7.2–3.
44. Seneca, *Natural Questions* 4a.2.7.
45. Cary and Warmington, *Ancient Explorers* 211–13; Thomson, *History* 272.
46. Ptolemy, *Geographical Guide* 1.9.
47. Ptolemy, *Geographical Guide* 1.17.6.
48. Ptolemy, *Geographical Guide* 4.7.26.
49. Richard Burton, however, believed it was a Greek translation of an indigenous name: Richard Francis Burton, *The Lake Regions of Central Equatorial Africa* (*Journal of the Royal Geographical Society* 29 [1859]) 205.
50. Cary and Warmington, *Ancient Explorers* 214–15.
51. Edward Rice, *Captain Sir Richard Francis Burton* (New York 1991) 355.
52. Strabo 4.5.2.
53. Strabo 4.5.3.
54. Suetonius, *Claudius* 17.1; Dio 59.25, 60.19–23.
55. Tacitus, *Agricola* 30.1.
56. Katherine Clarke, "An Island Nation: Re-Reading Tacitus' *Agricola*," *JRS* 91 (2001) 94–112.
57. Tacitus, *Agricola* 10, 29.2, 38.3; Dio 66.20.1.
58. Elmar Seebold, "Die Entdeckung der Orkneys in der Antike," *Glotta* 85 (2009) 195–216.
59. Tacitus, *Agricola* 4.
60. Stan Wolfson, *Tacitus, Thule and Caledonia: the Achievements of Agricola's Navy in their True Perspective* (Oxford 2008) 35–46.
61. Silius Italicus 3.597.
62. Freeman, *Ireland* 33–6.
63. Strabo 1.4.3; see also 2.1.13, 17–18; 4.5.5.
64. Caesar, *Gallic War* 5.13.
65. Pomponius Mela 3.53.
66. For a list, see Freeman, *Ireland* 65–84.
67. For Pomponius Mela, see F. E. Romer, *Pomponius Mela's Description of the World* (Ann Arbor, 1998); Kai Brodersen, *Terra Cognita* (*Spudasmata* 59, Zürich 1995) 87–94.
68. Pomponius Mela 2.96, 3.49.
69. Pomponius Mela 1.50, 3.38.
70. Romer, *Pomponius Mela's Description* 23.
71. Pomponius Mela 1.86, 3.90.
72. Pliny, *Natural History* 2.199, 3.132, 5.4, 6.5, 31, 199.
73. Pomponius Mela 3.45; see also Pliny, *Natural History* 2.170; Klaus Tausend, "Inder in Germanien," *OT* 5 (1999) 115–25.
74. Paul T. Keyser, "Turranius Gracilis," *EANS* 820.

75. Pliny, *Natural History* 2.89; 3.66.
76. Pliny, *Natural History* 2.5.
77. Pliny, *Natural History* 2.117–18.
78. Trevor Murphy, *Pliny the Elder's Natural History: The Empire in the Encyclopedia* (Oxford 2004) 129–64.
79. Sorcha Carey, *Pliny's Catalogue of Culture: Art and Empire in the Natural History* (Oxford 2003) 32–40.
80. Pliny, *Natural History, Preface* 1–5.
81. Molly Ayn Jones-Lewis, "Poison: Nature's Argument for the Roman Empire in Pliny the Elder's *Naturalis Historia*," *CW* 106 (2012) 51–74.
82. Pliny, *Natural History* 4.101, 105; 5.51–2; 36.1.
83. Pliny, *Natural History* 6.81–91.
84. Pliny, *Natural History* 3.46; A. H. M. Jones, *The Cities of the Eastern Roman Provinces* (Oxford 1937) 491–6.
85. Pliny, *Natural History* 6.178–80.
86. Plutarch, *Concerning the Face that Appears on the Globe of the Moon* 26.
87. Plutarch, *Symposium* 7.4.7.

Chapter 10 The Later Roman Empire

1. Krisztina Hoppál, "The Roman Empire According to the Ancient Chinese Sources," *AAntHung* 51 (2011) 263–306.
2. Ferguson, "China and Rome," 591–2.
3. Schoff, *Parthian Stations* 40–2.
4. *Supra*, pp. 165–6.
5. Cary and Warmington, *Ancient Explorers* 105–6.
6. D. D. Leslie and K. H. J. Gardiner, *The Roman Empire in Chinese Sources* (Rome 1996) 3–31.
7. *Periplous of the Erythraian Sea* 39, 49, 64.
8. Tacitus, *Annals* 2.33; Suetonius, *Gaius* 52.
9. Ptolemy, *Geographical Guide* 1.11–12.
10. Berggren and Jones, *Ptolemy's Geography* 179.
11. *BA*, Map 6, D2.
12. Berggren and Jones, *Ptolemy's Geography* 152.
13. Ferguson, "China and Rome," 594–5.
14. Casson, *Periplus* 27.
15. Pausanias 6.26.6.
16. Berggren and Jones, *Ptolemy's Geography* 23.
17. Alexander Jones, "Ptolemy's Geography: Mapmaking and the Scientific Enterprise," in *Ancient Perspectives: Maps and their Place in Mesopotamia, Egypt, Greece and Rome* (ed. Richard J. A. Talbert, Chicago 2012) 109–10.
18. Berggren and Jones, *Ptolemy's Geography* 17–18.

19. Ptolemy, *Geographical Guide* 1.6.
20. Ptolemy, *Geographical Guide* 1.4.
21. Richard J. A. Talbert. Review of Dueck and Brodersen, *Geography in Classical Antiquity: Key Themes in Ancient History* (Cambridge 2012), *BMCR* 2012.12.29.
22. Ptolemy, *Geographical Guide* 8.2; Berggren and Jones, *Ptolemy's Geography* 41–50.
23. Thomson, *History* 337–8.
24. For Columbus' use of the ancient sources, see V. Frederick Rickey, "How Columbus Encountered America," *Mathematics Magazine* 65 (1992) 219–25.
25. Samuel Eliot Morison, *Christopher Columbus, Mariner* (New York 1956) 18–19, 69.
26. Ptolemy, *Geographical Guide* 6.7.1–47, 7.4.1–13; Ptolemaios (ed. Stückelberger and Grasshoff) 874–5, 906–7.
27. Alastair Strang, "Explaining Ptolemy's Roman Britain," *Britannia* 38 (1997) 1–30; Barri Jones and Ian Keillar, "Marinus, Ptolemy and the Turning of Scotland," *Britannia* 27 (1996) 43–9; James J. Tierney, "Ptolemy's Map of Scotland," *JHS* 79 (1959) 132–48.
28. Ptolemy, *Geographical Guide* 7.5.4, 6.14.2.
29. Ptolemy, *Geographical Guide* 7.5.2; Hipparchos, *Against the "Geography" of Eratosthenes* F4 (= Strabo 1.1.9); Polybios 3.38.
30. Ptolemy, *Geographical Guide* 7.3.3; Thomson, *History* 277–8.
31. Cicero, *Letters to Atticus* #24, 26.
32. Cicero, *Republic* 6.21–2.
33. Cicero, *Academica* 2.123; Plutarch, *Concerning the Face that Appears on the Globe of the Moon* 26; Paul Coones, "The Geographical Significance of Plutarch's Dialogue, *Concerning the Face which Appears in the Orb of the Moon*," *Transactions of the Institute of British Geographers* 8 (1983) 361–72.
34. Ptolemy, *Geographical Guide* 1.7.11, 3.5.3.
35. Ptolemy, *Geographical Guide* 2.11.33, 35.
36. Ptolemy, *Geographical Guide* 3.5.1; this was an ethnym known as early as Herodotos (4.21).
37. *Periplous of the Erythraian Sea* 66.
38. Wheeler, *Rome Beyond* 28, 47, 98; Thomas Grane, "Did the Romans Really Know (or Care) about Southern Scandinavia? An Archaeological Perspective," in *Beyond the Roman Frontier: Roman Influences on Northern Barbaricum* (ed. Thomas Grane, Rome 2007) 7–29.
39. J. M. Alonso-Nuñez, "A Note on Roman Coins Found in Iceland," *OJA* 5 (1986) 121–2; M. P. Charlesworth, "A Roman Imperial Coin From Nairobi," *NC* 9 (1949) 107–10; Yves Janvier, "La geographie greco-romaine a-t-elle connu Madagascar?" *Omaly sy Anio* 1–2 (1975) 11–41.

40. Wheeler, *Rome Beyond* 203–6.
41. *Supra*, p. 153.
42. Athenaios 3.121a, 14.657f.
43. See further, *infra*, p. 216.
44. L. Richardson jr, *A New Topographical Dictionary of Ancient Rome* (Baltimore 1992) 319–20.
45. Dilke, *Greek and Roman Maps* 104–6.
46. Richard J. A. Talbert, *Rome's World: The Peutinger Map Reconsidered* (Cambridge 2010).
47. Talbert, *Rome's World* 144–6.
48. Tønnes Bekker-Nielsen, "Terra Incognita: the Subjective Geography of the Roman Empire," in *Studies in Ancient History and Numismatics Presented to Rudi Thomsen* (ed. Erik Christiansen *et al.*, Aarhus 1988) 148–61.
49. Strabo 1.1.1.
50. Herodotos 2.104–6.
51. Strabo 16.2.35–9; Cinzia Achille, "Strabone e la storia giudaica: la progressiva corruzione della legge di Mosè," *Sungraphe* 6 (2004) 89–105.
52. Augustine, *City of God* 8.11, 16.9.
53. John Watson McCrindle, *The Christian Topography of Cosmas, an Egyptian Monk* (London 1897).
54. Kosmas 113.
55. Kosmas 132, 140.
56. *Periplous of the Erythraian Sea* 4; Stanley M. Burstein, "Axum and the Fall of Meroe," *JARCE* 18 (1981) 47–50.
57. Kosmas 134–5.
58. Kosmas 265–7.
59. Kosmas 149.
60. Kosmas 194.
61. For the printing history of these works, see *Dictionary of Greek and Latin Authors and Texts, BNP Supplement* 2 (ed. Manfred Landfester *et al.*, 2009) 505, 548–9, 598–9.

Appendix 2 Some Further Notes on Mapping in Antiquity

1. Francesca Rochberg, "The Expression of Terrestrial and Celestial Order in Ancient Mesopotamia," in *Ancient Perspectives: Maps and Their Place in Mesopotamia, Egypt, Greece and Rome* (ed. Richard J. A. Talbert, Chicago 2012) 9–46.
2. For the terminology involved, see Dilke, *Greek and Roman Maps* 196–7.
3. Ephoros F30a (= Strabo 1.2.28).

4. Kosmas 149; Geus, "Space and Geography," 233.
5. Stanley M. Burstein, Review of Kai Brodersen and Jas Elsner (eds), *Images and Texts on the "Artemidorus Papyrus," BMCR* 2010.10.20; Richard Janko, "The Artemidorus Papyrus," *CR* 59 (2009) 403–10.
6. Richard J. A. Talbert, "P. Artemid.: The Map," in *Images and Texts on the "Artemidorus Papyrus"* (ed. Kai Brodersen and Jas Elsner, Stuttgart 2009) 57–64.
7. Irby, "Mapping the World" 81–107.
8. Agathemeros 1.1; Diogenes Laertios 2.2.
9. Herodotos 5.49.
10. For example, Herodotos' description of the Persian Royal Road (5.52–4).
11. Herodotos 4.42–4.
12. Diogenes Laertios 9.46; Agathemeros 1.1–2; Plutarch, *On Isis and Osiris* 6.
13. Diogenes Laertios 5.51.
14. Stanley F. Bonner, *Education in Ancient Rome* (Berkeley 1977) 130–1. Maps that were used for teaching were a conspicuous feature of the schools of Augustodunum (modern Autun in France), a major education center in the Late Roman period (R. Martin, "Augustodunum," *PECS* 122–3).
15. Apollonios 4.279–81; the map must be interpreted as a construct from Apollonios' own era of the third century BC, not a document from early times.
16. Eratosthenes, *Geography* F25, 30 (= Strabo 1.4.1, 2.5.5–6); Roller, *Eratosthenes* 21.
17. Eratosthenes, *Geography* F51, 64 (= Strabo 2.1.10–11, 2.1.34).
18. Hipparchos, *Against the "Geography" of Eratosthenes* F32 (= Strabo 2.1.40).
19. Strabo 2.5.10; Thomson, *History* 217–18.
20. *Supra*, pp. 196–200.
21. Brodersen, *Terra Cognita*; Mattern, "Rome and the Enemy," 24–66.
22. Livy 41.28.8–10; see Sherk, "Roman Geographical Exploration," 558–60, for other examples.
23. *Supra*, pp. 166–7; Pliny, *Natural History* 3.17: "orbem terrarum urbi spectandus."
24. Tina Najbjerg and Jennifer Trimble, "The Severan Marble Plan Since 1960," in *Formae Urbis Romae* (ed. Roberto Meneghini and Riccardo Santangeli Valenzani, Rome 2006) 75–101.
25. *Supra*, p. 203; Talbert, "*Urbs Roma*" 163–91.
26. Propertius, *Elegies* (ed. G. P. Goold, Cambridge, Mass. 1990) 2.
27. Propertius 4.3.37–40.

BIBLIOGRAPHY

Abel, Karlhaus. "Zone," *RE Supp.* 14 (1974) 987–1188.

Achille, Cinzia. "Strabone e la storia giudaica: la progressiva corruzione della legge di Mosè," *Sungraphe* 6 (2004) 89–105.

Agatharchides. *On the Erythraean Sea* (trans. and ed. Stanley M. Burstein, London 1989).

Alonso-Nuñez, J. M. "A Note on Roman Coins Found in Iceland," *OJA* 5 (1986) 121–2.

Aristotle. *Meteorologica* (tr. H. D. P. Lee, Cambridge, Mass. 1952).

Arnaud, Pascal. "Texte et carte de Marcus Agrippa: historiographie et données textuelles," *Geographia antiqua* 16–17 (2007–8) 73–126.

Aujac, Germaine. "Les modes de representation du monde habité d'Aristote a Ptolémée," *AFM* 16 (1983) 14–19.

Avienus, Rufus Festus. *Ora maritima* (ed. J. P. Murphy, Chicago 1977).

Barber, G. L. *The Historian Ephorus* (Cambridge 1935).

Bekker-Nielsen, Tønnes. "Terra Incognita: the Subjective Geography of the Roman Empire," in *Studies in Ancient History and Numismatics Presented to Rudi Thomsen* (ed. Erik Christiansen *et al.*, Aarhus 1988) 148–61.

Beresford, James. *The Ancient Sailing Season* (Leiden 2013).

Berggren, J. Lennart and Alexander Jones. *Ptolemy's Geography: An Annotated Translation of the Theoretical Chapters* (Princeton 2000).

Bianchetti, Serena. "Il valore del racconto di viaggio nell'opera geografica di Eratosthene," in *Vermessing der Oikumene* (ed. Klaus Geus and Michael Rathmann, Berlin 2013) 77–86.

Bigwood, J. M. "Ctesias As Historian of the Persian Wars," *Phoenix* 32 (1978) 19–41.

Bili, Tomislav. "Crates of Mallos and Pytheas of Massalia: Examples of Homeric Exegesis in Terms of Mathematical Geography," *TAPA* 142 (2012) 295–328.

Blomqvist, Jerker. *The Date and Origin of the Greek Version of Hanno's Periplus* (Lund 1979).

Boardman, John. *The Greeks Overseas: Their Early Colonies and Trade* (new and enlarged edition, London 1980).

Bolton, J. D. P. *Aristeas of Proconnesus* (Oxford 1962).
Bonner, Stanley F. *Education in Ancient Rome* (Berkeley 1977).
Bosworth, A. B. *A Historical Commentary on Arrian's History of Alexander* (Oxford 1980–).
——— "The Historical Setting of Megasthenes' *Indica*," *CP* 91 (1996) 113–27.
Bowersock, G. W. *Roman Arabia* (Cambridge, MA, 1983).
——— "Perfumes and Power," in *Profumi d'Arabia* (ed. Alessandra Avanzini, Rome 1997) 543–56.
——— "The East–West Orientation of Mediterranean Studies and the Meaning of North and South in Antiquity," in *Rethinking the Mediterranean* (ed. W. V. Harris, Oxford 2005) 167–78.
Braund, David. *Georgia in Antiquity* (Oxford 1994).
Bridgman, Timothy P. *Hyperboreans: Myth and History in Celtic-Hellenic Contacts* (New York 2005).
Brinken, Anna-Dorothee von den. "Antipodes," in *Trade, Travel, and Exploration in the Middle Ages* (ed. John Block Friedman *et al.*, New York 2000) 27–9.
Brodersen, Kai. *Terra Cognita* (*Spudasmata* 59, Zürich 1995).
Brogan, Olwen. "Trade Between the Roman Empire and the Free Germans," *JRS* 26 (1936) 195–222.
Brown, Truesdell S. *Onesicritus* (Berkeley 1949).
Bunbury, E. H. *A History of Ancient Geography* (London 1883).
Burstein, Stanley M. "Axum and the Fall of Meroe," *JARCE* 18 (1981) 47–50.
——— *Commentary* to *BNJ* #666.
——— *Commentary* to *BNJ* #670.
——— Review of Kai Brodersen and Jas Elsner (eds), *Images and Texts on the "Artemidorus Papyrus," BMCR* 2010.10.20.
Burton, Richard Francis. *The Lake Regions of Central Equatorial Africa* (= *Journal of the Royal Geographical Society* 29 [1859]).
Carey, Sorcha. *Pliny's Catalogue of Culture: Art and Empire in the Natural History* (Oxford 2003).
Carpenter, Rhys. "Phoenicians in the West," *AJA* 62 (1958) 35–53.
——— *Beyond the Pillars of Herakles* (New York 1966).
Cary, M. and E. H. Warmington. *The Ancient Explorers* (Baltimore 1963).
Casson, Lionel. *Ships and Seamanship in the Ancient World* (Princeton 1971).
——— *The Periplus Maris Erythraei* (Princeton 1989).
——— "Ptolemy II and the Hunting of African Elephants," *TAPA* 123 (1993) 247–60.
Cavalli-Sforza, Luca. "Demographic Data," in *African Pygmies* (ed. Luigi Luca Cavalli-Sforza, Orlando 1986) 23–44.
Chapelle, F. de la. "L'expédition de Suetonius Paulinus dans le sud-est du Maroc," *Hespéris* 19 (1934) 107–24.
Charlesworth, M. P. "A Roman Imperial Coin From Nairobi," *NC* 9 (1949) 107–10.
Clarke, Katherine. *Between Geography and History: Hellenistic Constructions of the Roman World* (Oxford 1999).

——— "An Island Nation: Re-Reading Tacitus' 'Agricola'," *JRS* 91 (2001) 94–112.

Colonna, Giovanni. "I Greci di Adria," *RSA* 4 (1974) 1–21.

Coones, Paul. "The Geographical Significance of Plutarch's Dialogue, *Concerning the Face which Appears in the Orb of the Moon," Transactions of the Institute of British Geographers* 8 (1983) 361–72.

Cunliffe, Barry. *The Extraordinary Voyage of Pytheas the Greek* (London 2001).

——— *Facing the Ocean: The Atlantic and its Peoples, 8000 BC – AD 1500* (Oxford 2001).

Dalby, Andrew. *Food in the Ancient World From A to Z* (London 2003).

Dicks, D. R. *The Geographical Fragments of Hipparchus* (London 1960).

Dictionary of Greek and Latin Authors and Texts (*BNP Supplement* 2, ed. Manfred Landfester *et al.*, 2009).

Dietler, Michael. "Colonial Encounters in Iberia and the Western Mediterranean: An Exploratory Framework," in *Colonial Encounters in Ancient Iberia* (ed. Michael Dietler and Carolina López-Ruiz, Chicago 2009) 3–48.

Dilke, O. A. W. *Greek and Roman Maps* (Ithaca 1985).

Diller, Aubrey. "Geographical Latitudes in Eratosthenes, Hipparchus and Posidonius," *Klio* 27 (1934) 258–69.

——— "The Ancient Measurements of the Earth," *Isis* 40 (1949) 6–9.

——— "Agathemerus, *Sketch of Geography*," *GRBS* 16 (1975) 59–76.

Dueck, Daniela and Kai Brodersen. *Geography in Classical Antiquity: Key Themes in Ancient History* (Cambridge 2012).

Eichel, Marijean H. and Joan Markley Todd. "A Note on Polybius' Voyage to Africa in 146 BC," *CP* 71 (1976) 237–43.

Engels, Johannes. "Geography and History," in *A Companion to Greek and Roman Historiography* (ed. John Marincola, Malden, Mass. 2011) 541–52.

——— "Artemidoros of Ephesos and Strabo of Amasia," in *Intorno al Papiro di Artemidoro 2: Geografia e Cartografia* (ed. C. Gallazzi *et al.*, Rome 2012) 139–55.

——— "Kulturgeographie im Hellenismus: Die Rezeption des Eratosthenes und Poseidonios durch Strabon in den *Geographika*," in *Vermessing der Oikumene* (ed. Klaus Geus and Michael Rathmann, Berlin 2013) 87–99.

Evans, James and J. Lennart Berggren. *Geminos's Introduction to the Phenomena* (Princeton 2006).

Fabre-Serris, Jacqueline. "Comment parles des Amazones? L'exemple de Diodore de Sicile et de Strabon," *CRIPEL* 27 (2008) 39–48.

Fear, Andrew T. "Odysseus and Spain," *Prometheus* 18 (1992) 19–26.

Ferguson, John. "Classical Contacts with West Africa," in *Africa in Classical Antiquity* (ed. L. A. Thompson and John Ferguson, Ibadan 1969) 1–25.

——— "China and Rome," *ANRW* 2.9.2 (1978) 581–603.

Fraser, P. M. *Ptolemaic Alexandria* (Oxford 1972).

——— *Cities of Alexander the Great* (Oxford 1996).

Frazer, J. G. *Pausanias's Description of Greece* (reprint, New York 1965).

Freeman, Philip. *Ireland and the Classical World* (Austin 2001).
Frost, Frank J. "Voyages of the Imagination," *Archaeology* 46.2 (March/April 1993) 44–51.
Gardiner-Garden, John R. "Greek Conceptions on Inner Asian Geography and Ethnography From Ephoros to Eratosthenes," *Papers on Inner Asia* 9 (Bloomington, Ind., 1987).
Gawlikowski, Michal. "Thapsacus and Zeugma: The Crossing of the Euphrates in Antiquity," *Iraq* 58 (1996) 123–33.
Gehrke, Hans-Joachim. "Thukydides und die Geographie," *Geographia antiqua* 18 (2009) 133–43.
Geus, Klaus. *Eratosthenes von Kyrene* (Munich 2002).
——— "Measuring the Earth and the *Oikoumene*: Zones, Meridians, *Sphragides* and Some Other Geographical Terms Used by Eratosthenes of Cyrene," in *Space in the Roman World* (*Antike Kultur und Geschichte* 5, ed. Kai Brodersen, Münster 2004) 11–26.
——— "Space and Geography," in *A Companion to the Hellenistic World*" (ed. Andrew Erskine, Malden, Mass. 2005) 232–45.
Gill, David W. J. "Silver Anchors and Cargoes of Oil: Some Observations on Phoenician Trade in the Western Mediterranean," *BSR* 56 (1988) 1–12.
Goldstein, Bernard R. "Eratosthenes on the 'Measurement' of the Earth," *Historia Mathematica* 11 (1984) 411–16.
Gorman, Vanessa B. *Miletos: The Ornament of Ionia* (Ann Arbor 2001).
Graham, A. J. "The Colonial Expansion of Greece," *CAH* 3.3 (2nd edn 1982) 83–159.
Graham, Daniel W. *The Texts of Early Greek Philosophy* (Cambridge 2010).
Grane, Thomas. "Did the Romans Really Know (or Care) about Southern Scandinavia? An Archaeological Perspective," in *Beyond the Roman Frontier: Roman Influences on Northern Barbaricum* (ed. Thomas Grane, Rome 2007) 7–29.
Green, Peter. *The Argonautika; The Story of Jason and the Quest for the Golden Fleece* (Berkeley 1997).
Gurukkal, Rajan and Dick Whittaker. "In Search of Muziris," *JRA* 14 (2001) 335–50.
Hahn, Robert. *Anaximander and the Architects: The Contributions of Egyptian and Greek Architectural Technologies to the Origins of Greek Philosophy* (Albany 2001).
Halfmann, Helmut. *Itinera principum* (Stuttgart 1986).
Hamilton, J. R. "Alexander and the Aral," *CQ* 21 (1971) 106–11.
Harden, D. B. "The Phoenicians on the West Coast of Africa," *Antiquity* 22 (1948) 141–50.
Harrell, James A. "Discovery of the Red Sea Source of *Topazos* (Ancient Gem Periodot) on Zabargad Island, Egypt," in *Twelfth Annual Sikankas Symposium. Periodot and Uncommon Gem Minerals* (ed. Lisbet Thoresen, Fallbrook, Cal. 2014) 16–30.
Heckel, Waldmar. *Who's Who in the Age of Alexander the Great* (Oxford 2006).
Herzhoff, Bernard. "Lotus," *BNP* 7 (2005) 822–3.

Hind, John. "Pyrene and the Date of the 'Massaliot Sailing Manual," *RSA* 2 (1972) 39–52.
Hope Simpson, R. and J. F. Lazenby, *The Catalogue of Ships in Homer's Iliad* (Oxford 1970).
Hoppál, Krisztina. "The Roman Empire According to the Ancient Chinese Sources," *AAntHung* 51 (2011) 263–306.
Ilyushechkina, Ekaterina. "Das Weltbild des Dionysios Periegetes," in *Vermessing der Oikumene* (ed. Klaus Geus and Michael Rathmann, Berlin 2013) 137–61.
Irby, Georgia L. "Mapping the World: Greek Initiatives from Homer to Eratosthenes," in *Ancient Perspectives: Maps and their Place in Mesopotamia, Egypt, Greece and Rome* (ed. Richard J. A. Talbert, Chicago 2012) 81–107.
——— "Hydrology: Ocean, Rivers, and Other Waterways," in *A Companion to Science, Technology, and Medicine in Ancient Greece and Rome* (ed. Georgia L. Irby, London 2016) 181–96.
Janko, Richard. "The Artemidorus Papyrus," *CR* 59 (2009) 403–10.
Janvier, Yves. "La geographie greco-romaine a-t-elle connu Madagascar?," *Omaly sy Anio* 1–2 (1975) 11–41.
Jones, A. H. M. *The Cities of the Eastern Roman Provinces* (Oxford 1937).
Jones, Alexander. "Ptolemy's Geography: Mapmaking and the Scientific Enterprise," in *Ancient Perspectives: Maps and their Place in Mesopotamia, Egypt, Greece and Rome* (ed. Richard J. A. Talbert, Chicago 2012) 109–28.
Jones, Barri and Ian Keillar. "Marinus, Ptolemy and the Turning of Scotland," *Britannia* 27 (1996) 43–9.
Jones-Lewis, Molly Ayn. "Poison: Nature's Argument for the Roman Empire in Pliny the Elder's *Naturalis Historia*," *CW* 106 (2012) 51–74.
Jouanna, Jacques. *Hippocrate* 2.2: *Airs-Eaux-Lieux* (Paris 2003).
Karttunen, Klaus. *India in Early Greek Literature* (Helsinki 1989).
——— *India and the Hellenistic World* (Helsinki 1997).
Keyser, Paul T. "From Myth to Map: The Blessed Isles in the First Century BC," *AncW* 24 (1993) 149–67.
——— "The Geographical Work of Dikaiarchos," in *Dicaearchus of Messana: Text, Translation, and Discussion* (ed. William W. Fortenbaugh and Eckart Schütrumpf, New Brunswick 2001) 353–72.
——— "Turranius Gracilis," *EANS* 820.
——— and Georgia L. Irby-Massie (eds). *The Encyclopedia of Ancient Natural Scientists* (London 2008).
Kidd, I. G. *Posidonius* 2: *The Commentary* (Cambridge 1988).
——— *Posidonius* 3: *The Translation of the Fragments* (Cambridge 1999).
Kim, Lawrence. "The Portrait of Homer in Strabo's Geography," *CP* 102 (2007) 363–8.
——— *Homer Between History and Fiction in Imperial Greek Literature* (Cambridge 2010).
Kowalski, Jean-Marie. *Navigation et géographie dans l'antiquité gréco-romaine* (Paris 2012).

Krates of Mallos. *Cratete di Mallo, I frammenti* (ed. Maria Broggiato, La Spezia, 2001).
Lasserre, François. *Die Fragmente des Eudoxos von Knidos* (Berlin 1966).
——— "Strabon devant l'Empire romain," *ANRW* 30 (1982–3) 867–96.
Lattimore, Richmond. *The Iliad of Homer* (Chicago 1951).
Law, R. C. C. "The Garamantes and Trans-Saharan Enterprise in Classical Times," *Journal of African History* 8 (1967) 181–200.
Leslie, D. W. and K. H. J. Gardiner. *The Roman Empire in Chinese Sources* (Rome 1996).
Levick, Barbara. *Tiberius the Politician* (London 1976).
Lieberman, Samuel. "Who Were Pliny's Blue-Eyed Chinese?," *CP* 52 (1957) 174–7.
Llewellyn-Jones, Lloyd and James Robson. *Ctesias' History of Persia* (London 2010).
López-Ruiz, Carolina. "Tarshish and Tartessos Revisited: Textual Problems and Historical Implications," in *Colonial Encounters in Ancient Iberia* (ed. Michael Dietler and Carolina López-Ruiz, Chicago 2009) 255–80.
Lupher, David A. *Romans in a New World* (Ann Arbor 2003).
McCrindle, John Watson. *The Christian Topography of Cosmas, an Egyptian Monk* (London 1897).
Martin, R. "Augustodunum," *PECS* 122–3.
Mattern, Susan P. *Rome and the Enemy: Imperial Strategy in the Principate* (Berkeley 1999).
Mendell, Henry. "Eudoxos of Knidos," *EANS* 310–13.
Millar, Fergus. "Caravan Cities: The Roman Near East and Long-Distance Trade By Land," in *Essays in Honor of Geoffrey Rickman* (ed. Michael M. Austin *et al.*, London 1998) 119–37.
Miller, J. Innes. *The Spice Trade of the Roman Empire* (Oxford 1969).
Morison, Samuel Eliot. *Christopher Columbus, Mariner* (New York 1956).
Muhly, James D. "Sources of Tin and the Beginnings of Bronze Metallurgy," *AJA* 89 (1985) 275–91.
Murphy, Trevor. *Pliny the Elder's Natural History: The Empire in the Encyclopedia* (Oxford 2004).
Najbjerg, Tina and Jennifer Trimble. "The Severan Marble Plan Since 1960," in *Formae Urbis Romae* (ed. Roberto Meneghini and Riccardo Santangeli Valenzani, Rome 2006) 75–101.
Nansen, Fridtjof. *In Northern Mists: Arctic Exploration in Early Times* (tr. Arthur G. Chater, New York 1911).
Nappo, Dario. "On the Location of Leuke Kome," *JRA* 23 (2010) 335–48.
Nicholls, R. V. "Recent Aquisitions by the Fitzwilliam Museum, Cambridge," *AR* 12 (1965–6) 44–51.
Nicolet, Claude. *Space, Geography, and Politics* (Ann Arbor 1991).
Niemeyr, Hans Georg *et al.* "Phoenicians, Poeni," *BNP* 11 (2007) 148–69.
Parker, Grant. *The Making of Roman India* (Cambridge 2008).

Parker, Victor. *Commentary to BNJ #70*.
Patterson, Lee E. "Geographers as Mythographers: The Case of Strabo," in *Writing Myth: Mythography in the Ancient World* (ed. Stephan M. Trzaskoma and R. Scott Smith, Leuven 2013) 201–21.
Pearson, Lionel. *Early Ionian Historians* (Oxford 1939).
——— *The Lost Histories of Alexander the Great* (New York 1960).
——— *The Local Historians of Attica* (Atlanta 1981).
Phillips, E. D. "Odysseus in Italy," *JHS* 73 (1953) 53–67.
Pliny. *Histoire Naturelle* 6, part 2 (ed. Jacques André and Jean Filliozat, Paris 2003).
Polverini, Leandro. "Cesare e la geografia," *Semanas de estudios romanos* 14 (2005) 59–72.
Potts, D. T. *The Arabian Gulf in Antiquity* 2: *From Alexander the Great to the Coming of Islam* (Oxford 1990).
Prontera, Francesco. *Geografia e storia nella Grecia antica* (Florence 2011).
Propertius. *Elegies* (ed. G. P. Goold, Cambridge, Mass. 1990).
Ptolemaios. *Handbuch der Geographie* (ed. Alfred Stückelberger and Gerd Grasshoff, Basel 2006).
Pytheas. *L'Oceano* (ed. Serena Bianchetti, Pisa 1998).
Raschke, Manfred G. "New Studies in Roman Commerce with the East," *ANRW* 2.9 (1978) 604–1361.
Rawlinson, George. *The History of Herodotus* (New York 1860–2).
Rice, Edward. *Captain Sir Richard Francis Burton* (New York 1991).
Richardson Jr, L. *A New Topographical Dictionary of Ancient Rome* (Baltimore 1992).
Rickey, V. Frederick. "How Columbus Encountered America," *Mathematics Magazine* 65 (1992) 219–25.
Rochberg, Francesca. "The Expression of Terrestrial and Celestial Order in Ancient Mesopotamia," in *Ancient Perspectives: Maps and their Place in Mesopotamia, Egypt, Greece and Rome* (ed. Richard J. A. Talbert, Chicago 2012) 9–46.
Roller, Duane W. "Columns in Stone: Anaximandros' Conception of the World," *AntCl* 53 (1989) 185–9.
——— *The World of Juba II and Kleopatra Selene: Royal Scholarship on Rome's African Frontier* (London 2003).
——— *Scholarly Kings: The Writings of Juba II of Mauretania, Archelaos of Kappadokia, Herod the Great, and the Emperor Claudius* (Chicago 2004).
——— "Seleukos of Seleukeia," *AntCl* 74 (2005) 111–18.
——— *Through the Pillars of Herakles: Greco-Roman Exploration of the Atlantic* (London 2006).
——— *Eratosthenes' Geography* (Princeton 2010).
——— *The Geography of Strabo* (Cambridge 2014).
Romer, F. E. *Pomponius Mela's Description of the World* (Ann Arbor 1998).
Romm, James S. *The Edges of the Earth in Ancient Thought* (Princeton 1992).
Roseman, Christina Horst. *Pytheas of Massalia: On the Ocean* (Chicago 1994).

Rostovtzeff, M. *Social and Economic History of the Hellenistic World* (Oxford 1941).
Salama, P. "The Sahara in Classical Antiquity," in *General History of Africa 2: Ancient Civilizations of Africa* (ed. G. Mokhtar, Paris and London 1981) 513–32.
Salles, Jean-François. "Achaemenid and Hellenistic Trade in the Indian Ocean," in *The Indian Ocean in Antiquity* (ed. Julian Reade, London 1996) 251–67.
Scheckley, R. "Romans In Rio?," *Omni* 5 (June 1983) 43.
Schoff, Wilfred H. *Parthian Stations of Isidore of Charax* (London 1914).
Scodel, Ruth. "The Paths of Day and Night," *Ordia Prima* 2 (2003) 83–6.
Scullard, H. H. *From the Gracchi to Nero* (4th edn, London 1976).
Seebold, Elmar. "Die Entdeckung der Orkneys in der Antike," *Glotta* 85 (2009) 195–216.
Sherk, Robert K. "Roman Geographical Exploration and Military Maps," *ANRW* 2.1 (1974) 534–62.
Shipley, Graham. *Pseudo-Skylax's Periplous: The Circumnavigation of the Inhabited World* (Exeter 2011).
Sidebotham, Steven E. "Ports of the Red Sea and the Arabia-India Trade," in *The Eastern Frontier of the Roman Empire* (ed. D. H. French and C. S. Lightfoot, *BAR-IS* 553, 1989) 485–513.
——— *Berenike and the Ancient Maritime Spice Route* (Berkeley 2011).
Speke, John Hanning. *Journal of the Discovery of the Source of the Nile* (New York 1868).
Standish, J. F. "The Caspian Gates," *G&R* 17 (1970) 17–24.
Stiehle, R. "Der Geograph Artemidoros von Ephesos," *Philologus* 11 (1856) 193–244.
Strang, Alastair. "Explaining Ptolemy's Roman Britain," *Britannia* 38 (1997) 1–30.
Stubbings, Frank H. "The Recession of Mycenaean Civilization," *CAH* 2.2 (3rd edn, Cambridge 1975) 338–58.
Sullivan, Richard D. "Dynasts in Pontus," *ANRW* 7.2 (1980) 913–30.
Syme, Ronald. *Anatolica: Studies in Strabo* (ed. Anthony Birley, Oxford 1995).
Talbert, Richard J. A. "P. Artemid.: The Map," in *Images and Texts on the "Artemidorus Papyrus"* (ed. Kai Brodersen and Jas Elsner, Stuttgart 2009) 57–64.
——— *Rome's World: The Peutinger Map Reconsidered* (Cambridge 2010).
——— Review of Dueck and Brodersen, *Geography in Classical Antiquity: Key Themes in Ancient History* (Cambridge 2012), *BMCR* 2012.12.29.
——— "*Urbs Roma* to *Orbis Romanus*: Roman Mapping on the Grand Scale," in *Ancient Perspectives: Maps and their Place in Mesopotamia, Egypt, Greece and Rome* (ed. Richard J. A. Talbert, Chicago 2012) 163–91.
Tarn, W. W. *The Greeks in Bactria and India* (revised 3rd edn, Chicago 1997).

Tausend, Klaus. "Inder in Germanien," *OT* 5 (1999) 115–25.
Taylour, Lord William. *Mycenaean Pottery in Italy and Adjacent Areas* (Cambridge 1955).
Thiel, J. H. *Eudoxus of Cyzicus* (Groningen 1939).
Thompson, L. A. "Eastern Africa and the Graeco-Roman World (to A. D. 641)," in *Africa in Classical Antiquity* (ed. L. A. Thompson and John Ferguson, Ibadan 1969) 26–61.
Thomson, J. Oliver. *History of Ancient Geography* (Cambridge 1948).
Tierney, James J. "Ptolemy's Map of Scotland," *JHS* 79 (1959) 132–48.
——— "The Celtic Ethnography of Posidonius," *Proceedings of the Royal Irish Academy* 60C5 (1960) 189–275.
——— "The Map of Agrippa," *Proceedings of the Royal Irish Academy* 63C (1963) 151–66.
Van Beek, Gus W. "Frankincense and Myrrh in Ancient South Arabia," *JAOS* 78 (1958) 141–51.
Vermeule, Emily. *Greece in the Bronze Age* (Chicago 1964).
Wagner, Emil August. *Die Erdbeschreibung des Timosthenes von Rhodus* (Leipzig 1888).
Walbank, F. W. "The Geography of Polybius," *ClMed* 9 (1947) 155–82.
——— *Polybius* (Berkeley 1972).
Wallace, Paul. *Strabo's Description of Boiotia: A Commentary* (Heidelberg 1979).
Warmington, B. H. *Carthage* (revised edition, New York 1969).
Wenskus, Reinhard. "Pytheas und der Bernsteinhandel," in *Untersuchungen zu Handel und Verkehr der vor- und frühgeschichtlichen Zeit in Mittel- und Nordeuropa* (ed. Klaus Duwel *et al.*, Göttingen 1985) 84–108.
West, Stephanie. "'The Most Marvellous of All Seas': The Greek Encounter With the Euxine," *G* 50 (2003) 151–67.
Wheeler, Mortimer. *Rome Beyond the Imperial Frontiers* (London 1955).
Wolfson, Stan. *Tacitus, Thule and Caledonia: the Achievements of Agricola's Navy in their True Perspective* (Oxford 2008).
Xenophanes of Colophon. *Fragments* (ed. J. H. Lesher, Toronto 1992).

INDEX OF PASSAGES CITED

Greek and Latin Authors

Biblical Citations

Inscriptions

GENERAL INDEX

Literary figures whose works are quoted in the text are not included in the index (see Index of Passages Cited), unless their careers or writings are actually part of the discussion. The hometown of a cultural figure that forms part of the effective personal name (e.g. Eudoxos of Kyzikos) is not indexed separately unless it has some other significance. Ethnyms may be indexed with their toponyms.